An American Experience in Indonesia

A Publication of the
Center for Developmental Change
University of Kentucky

An American Experience in Indonesia: The University of Kentucky Affiliation with the Agricultural University at Bogor

Howard W. Beers

The University Press of Kentucky

Standard Book Number: 8131-1235-4

Library of Congress Catalog Card Number: 75-132824

Copyright © 1971 by The University Press of Kentucky

A statewide cooperative scholarly publishing agency
serving Berea College, Centre College of Kentucky,
Eastern Kentucky University, Kentucky State College,
Morehead State University, Murray State University,
University of Kentucky, University of Louisville,
and Western Kentucky University.

Editorial and Sales Offices: Lexington, Kentucky 40506

Contents

Foreword

Broadening scope, growing size, and rising costs have brought American universities close to, and in many categories beyond, the limits of support from their usual sources. Educational programs have accordingly been somewhat constrained by community interest in their own cities and regions. To reach across oceans with costly educational activity in developing countries is possible only with support from unusual resources: foreign aid, foundation sponsorship, and international business interest. Because of its localistic relevance and its experience in community service, a university in the American interior has special fitness for technical assistance abroad, but reliance on local financing does not give it strength to become internationally involved. Not until the Truman administration innovated the Point Four concept of technical assistance and started the funding of the United States technical cooperation abroad did American universities go beyond their earlier small-scale ventures in international exchange programs to undertake major projects overseas.

Contracting with a United States government agency—successively the Economic Cooperation Administration, the Mutual Security Agency, the International Cooperation Administration, and the Agency for International Development—universities entered remote arenas of development at the close of World War II. Groups were assigned to work overseas at designated colleges or universities, or in government ministries, or at institutions for research or public service in countries that a few years earlier were only names on a map for most Americans. The universities and the specialists they sent abroad were competent, experienced, learned, and skillful within their native sociocultural habitats. At new and foreign sites, however, they had to temper their confidence with realistic humility in devising adaptations rather than simply transferring (as they had first thought might be possible) their expertise to developing agencies, institutions, and colleagues.

The University of Kentucky's first project was in counselling the government of Guatemala in agricultural science and technology. University personnel were posted in that Central American country between 1958 and 1963. A university-wide committee of faculty members and administrators (it was not yet the practice to have students on university committees) was pondering in 1955 and 1956 whether the university should become further involved and in other parts of the world. Both Dr. Beers and I served on that committee and took part in discussions about possible undertakings in Turkey and Yugoslavia. The work in Turkey was assigned by Washington to another university. Feasibility observers from Kentucky did, however, go to Yugoslavia for preliminary negotiations. It was a matter of surprise and growing excitement on the campus that somewhat ethnocentric Kentucky would even consider penetrating the iron curtain in those days. But the committee finally concluded that Yugoslavia would welcome only U.S. facilities and financing at that time and would accept American professors grudgingly as inevitable but unwelcome components in a package of aid.

When Washington mentioned Indonesia, the University of Kentucky committee responded with evident though initially circumspect interest. Feasibility observers from Lexington went to Djakarta in 1956; Professor Dr. Tojib Hadiwidjaja of Bogor, then an Eisenhower Fellow, supported energetically the idea of American university projects. Prime Minister Djuanda, with President Sukarno's approval, and U.S. Ambassador Howard P. Jones agreed on general terms by which science and American technology might be brought to bear upon economic development in the newly independent nation.

So the University of Kentucky was sent into service at the cities of Bandung and Bogor. In Bandung, the engineering sciences began development at ITB (Institut Teknologi, Bandung) soon after the separation of that institute from the University of Indonesia at Djakarta. The agricultural sciences were to be developed at Bogor, initially with the two faculties (colleges) in that city, which were still officially units of the University of Indonesia and which in 1963 became the autonomous IPB (Institut Pertanian, Bogor). The Kentucky Contract Team (Kenteam for short) worked in Bogor from late 1957 until early 1966. Its experience there is the subject of this book.

The need to link its work overseas with units and activities on campus was one factor leading eventually, but late in the life of the Bogor project, to the university's organization of the Center for Developmental Change. Initially, campus direction of the international projects was a responsibility of the Kentucky Research Foundation, a separately incorporated unit which had been brought into being by the university to receive and disburse funds for special purposes. A weakness in this arrangement was the absence of college and departmental responsibility and involvement. Legal and financial accountability was provided but not educational programming. A corrective effort was made to vest directorship of the Kentucky Research Foundation and the deanship of the graduate school in one person, thus drawing the international projects nominally into the latter educational subsystem of the university. A further adjustment was the organization of an Office of Overseas Programs which reported to the executive vice-president of the university. The next major rearrangement was accomplished only at the end of the Bogor project in the transfer of responsibility to the Center for Developmental Change, formed expressly to stimulate and facilitate the involvement of the university faculty and departments in developmental effort at home and abroad.

Perhaps if the Center for Developmental Change had been established before rather than after the university's work at Bogor had been undertaken, there might have been from the beginning more recourse to principles of institution-building and more concern with strategies of technical assistance and procedures for evaluation. But CDC's first director arrived on campus just one month before Kenteam left Bogor to come home, and the center's first involvement in a project of this type was in planning for a team to work in Thailand.

Several policy views and position statements were formulated to express the university's interest, based at least partly on Indonesian experiences. These guidelines enjoy particular significance, having been prepared by faculty committees that CDC organized for this purpose. They are now regularly applied in any review of project proposals at the University of Kentucky. The key policy now is that "overseas service activities must be so planned as to contribute systematically to the search for knowledge about the process of change and to improve international competence in the curriculum and teaching of the University."[1]

The guidelines are stated in terms of project characteristics, supports and university relevance. They require assurance that the project is feasible, that support is committed, that the university has requisite staff resources and interest, and that the anticipated duration will permit significant achievement. Moreover, a developmental project is required to have three interdependent phases: building an institution and training its staff, planning and initiating its program, and making it an agent of change within the groups it serves. Respect for intercultural differences must be reflected in each project's planning and operation. Research, with staff assured, is expected to include systematic study of such related problems as diffusion, regional development, cross-cultural cooperation, and training methods. Responsibility is acknowledged to assist in development of new organizational structure, personnel policies, and administration procedures. Cooperating staff from the host country, available in quality and quantity and appointed on schedule are required. Provision is made for training staff members of the host institution in three steps: first, a period of work in the host country with a university group member; next, a U.S. or "third-country" training period; and finally a post-return period of joint work in their home country by the trainees and the Kentucky team members.

The staffing of overseas teams is expected to include

[1] From "International Development Projects: Guidelines Interpreting University Policy" (Lexington: University of Kentucky, Center for Developmental Change, 1967).

graduate students assigned to do research and other project tasks under the supervision of their professors. A full-time position of chief of party is provided whenever the team exceeds five members. A home-campus dimension is required in each project. A number of additional guidelines concern details of administrative and logistical support, including staff and facilities; home campus personnel to cover recruiting, orientation, and programming; university control over assignment of faculty; adequate autonomy and operating flexibility; availability of funds for a working library; reasonable provision for travel; and cooperative short-term interchange of staff members among different university projects. The university learned from its experience in Guatemala and Bogor that these policy guidelines are needed; CDC administered their formulation in 1966, and determines from time to time whether they need revision.

Another set of guidelines, not expressing university policy but outlining strategies of developmental change, may be used in partial *ex post facto* evaluation of Kenteam's work. These are an accumulation of precepts and rules-of-thumb for technical assistance, "neither standardized nor codified, but . . . scattered in the practice of organizers, social workers, applied anthropologists, extension sociologists, community workers, therapists, arbitrators, negotiators, advertisers, managers, administrators and counsellors."[2] They may be understood as substrategies and tactics of "the group approach," which is a grand strategy of developmental change. Applied by developmental personnel (change agents) in their work with developing communities, these procedures are suitable for application also in institution-building, with the institution taking the place of the community.

One group of these precepts refers to the change agent's address or approach to the developmental task. They prescribe cultural congruity, or working within the cultural framework; working in consistency with the past experience of the host community; formulating procedures on the basis

[2] Howard W. Beers, "The Group-Approach to Developmental Change," in *Sociology of Underdevelopment*, ed. by Carle C. Zimmerman and Richard E. DuWors (Toronto: Copp Clark Publishing Co., 1970).

of mutual understanding, mutual respect, and feelings of need; and starting with the situation as initially given. Other strategies identify the roles of personnel. They include injunctions to work with both the formal and the informal leaders of the developing community (institution), with both innovators and others less ready to accept new practices.

Other prescriptions are principles of operation: involvement, "learning to do," demonstration, self-study, self-help, strengthening self-confidence, teaching positively, following a "natural" rather than a forced pace, utilizing existing organizations, asserting the dignity of work, retaining flexibility, sticking to feasible objectives and tasks, evaluating and applying the findings of evaluation in revising procedures, accepting "progress" in the adoption of innovations by stages, using different sources of information as appropriate, procuring requisite facilities, and helping agencies of government to organize for service to the developing community or institution.

A final group of principles includes strategies of result or outcome. They enjoin the change agent to avoid dissipating resources in controversy and conflict, to guard against producing anomie through destruction of norms, to accept realistic expectations, to acknowledge that solving old problems makes new ones, to assure that consequences of early effort will be beneficial, to transfer controls constructively, and to fulfill promises.

The impression of readers, after reading the chapters that follow, may be that Kenteam made many applications of these principles without ever spelling them out explicitly in plans of work. This conclusion will stand even though there were wide differences between Kenteam members in their sensitivity to educational process and teaching technique and even though their habits of work were not influenced enough by special knowledge about the principles of procedure in developmental change.

Too little study has been made of the institution-building process carried out within the context of interuniversity contractual affiliations. A major start has been made by the

CIC-AID rural development research project,[3] initiated world-wide during Kentucky's next-to-last year at Bogor. It was not feasible for CIC-AID representatives to visit the University of Kentucky team in the field. Data from the Indonesian case were included only tangentially, if at all, in the survey analysis, although the CIC-AID survey questions included some borrowed from the questionnaires Kenteam used at Bogor. The survey report, available shortly before the publication of this book, is a background against which the Bogor story can be viewed as a case-figure in the foreground. The documents are clearly intersupportive in their conclusions: the Bogor case confirms and reinforces by specific illustration each of the recommendations from the survey.

Ten recommendations emerged from the CIC-AID study, the first being for stronger commitment by all parties concerned to expanded and long-term programs. Clearly the time-span initially projected at Bogor and revised from time to time fostered neither the needed clarification of long-time goals nor the sequences of action by which goals could be achieved. In retrospect, one finds that the goals which were in mind at any moment were really instrumental for that year only, not ultimately for the decade or the quarter century. The limitation of appropriations and funding to one or two years at a time tends to inhibit completely longer-term planning, though it need not.

The second recommendation is for more flexible project agreements and improved AID-university liaison. The third calls for more research on the institution-building process and better utilization of knowledge in hand. The fourth asserts the relevance of the basic ideas, "if properly understood and employed," underlying the land-grant type of institution. Other recommendations call for "wider participation" (by host and U.S. personnel) in project planning and review; for change in factors which elicit negative attitudes from university personnel; for changes to improve orientation for

[3] See CIC-AID Rural Development Research Project, *Building Institutions to Serve Agriculture* (Lafayette, Indiana: Purdue University, Committee on Institutional Cooperation, 1968).

service abroad; and for better-planned and better-supported participant training programs. The ninth recommendation calls on universities to promote public understanding of international technical assistance. Finally, in a tenth recommendation, the study acknowledges the need for AID-university cooperation in strengthening the international competencies of universities. As this introduction is written, committees of government and university representatives are studying these recommendations and planning for their implementation.

Page by page this story about Kenteam at Bogor confirms and reinforces the CIC-AID findings. Or perhaps the inversion would be more accurate, since this book was written before the report. The involvement of Kentucky's professors shows all the needs for which the recommendations seek satisfaction: commitment, broadened effort, persistence, flexibility, liaison, evaluation, adaptation, "jointness," removal of barriers, orientation, cross-cultural training, developing public support, and strengthening university capacities.

The professors studied here earn passing grades with respect to each of several important qualities—team composition, competence, organizational skill, tact, understanding, directness, teaching methods, neutrality, sense of mission, empathy, utilization of abilities. If all the conditions sought by the survey's recommendations had prevailed when the Kentucky-AID contract was signed, Kenteam might have emerged from its institution-building labors with higher marks. But there are usually more than one set of "grade points." An Indonesian colleague, reviewing and analyzing the period of UK-IPB affiliation, would be looking for insights of developmental importance to his own country. The interest of an American guest professor is similar but not identical. The American wants to ask, "How effectively did we serve and what did we learn about how to take part in international education?" An Indonesian colleague might first ask, "Has the affiliation made us better able to serve the development of our nation?" Hence, Indonesian readers might seek sharper focus on their urgent concerns, on "more Indonesian" conclusions.

An Indonesian would have written this book differently,

so the author, endeavoring to compensate for ethnic error, has introduced the judgments of Indonesian colleagues to the extent that they were expressed in the surveys made. He compared the views of Americans and Indonesians by determining for each group its self-image and the image ascribed by the other. Greater statistical strength would have made comparisons firmer but they are usefully suggestive in the format used here.

The author gained the data for this book from several sources—the files of correspondence and documents, reports, numerous notes, and especially responses to special questionnaires. There were seven of these last, and they included 230 question units and many more subquestions. First, members of the IPB staff, both guest (Kenteam, 80% return) and host (a select panel of thirty-two members), were asked about their attitudes and experiences in connection with the affiliation. Second, Kenteam members were asked further about their own roles (80% return). Third, participants (staff members who received foreign training) were asked about their experience (33% sample). The wives of Kenteam members were asked about their own roles and about their experiences with Indonesian culture (73% return). Supervisors of participants who studied in American universities were asked several questions (60% return). Finally, a survey was made of documented recommendations to see what fate befell them in the affiliation.

Better, of course, had all persons queried made replies, but the schedule of questions (for both guest and host and in both languages) was ready for use at a time of maximum sensitivity in relations between host staff and guest staff (December 1965); it could be used only selectively and with consideration for the anxieties, fears, and uncertainties of the moment. This was a post-coup period when many Indonesians suspected even their friends and when political forces were seething, largely out of the view of Americans. The intensity of vital and immediate concerns, the urgency of preserving personal security, the effort of the leadership of the institute to keep it at work and on target relegated a very low priority to the questionnaire. The host-staff respondents, either by interview or delivered return, however, included all present and former deans, rectors, associate rectors, or presidium members then living in Indonesia. The thirty-two Indonesian respondents are not statistically a representative sample, but they are thoughtful persons and their responses identify prominent features of the affiliation even though they provide no measurement of pervasiveness or priority among those features.

The author of this book has written as a participant observer. He was a member of the Bogor team and chief of party from September 1962 through June 1966. Throughout that period I was executive vice-president of the University of Kentucky and was involved in the project as an interested educational programmer and administrator and as official visitor to the team in Indonesia in 1963 and 1965.

The author's association with the Bogor work of the university involves a quirk or two worthy to be explained. Actually his membership on the team postdates by nearly two years that of his wife, Bernice V. Beers, who served as an "on-post" specialist in the institute's department of family welfare (a better translation than "home economics" of ilmu kesedjahtaran keluarga). An interest which Bernice and Howard Beers shared in foreign service had led him to take indefinite leave from the University of Kentucky in 1959 to become an associate in community development on the field staff of the Agricultural Development Council (then known as the Council on Economic and Cultural Affairs). Having served as a professor of rural sociology at the University of Kentucky for twenty years and as department head for eleven, Dr. Beers was placed in Bogor by ADC to serve as a visiting professor of rural sociology on the faculty there and to represent ADC in promoting generally the contributions of the social sciences to rural development in Indonesia. In previous years, during periods of leave from the university, Professor Beers had promoted rural sociology, community development, and program evaluation in Greece under Near East Foundation and Fulbright sponsorship in 1949–1950; in the countries of Western Europe in 1955–1956 through the Organization for European Economic Cooperation, European Productivity

Agency; and in India through Ford Foundation sponsorship in 1958–1959.

It was an unusual and fortuitous occurrence that the University of Kentucky, by its contract with AID, and the Agricultural Development Council were represented by the same specialist, on leave from one to serve the other and in either case to do the same job at the same spot. After Beers had lived two years at Bogor in service for ADC, the University of Kentucky and the council arranged for him to reverse lines of responsibility again by taking leave from the position of ADC associate and resuming duty status for the University of Kentucky as chief of party on the team at the institute. Returning to on-campus duty in 1966, Beers was appointed to the staff of the Center for Developmental Change, becoming director in 1967.

This book is written then by the team's third and last chief of party, whose familiarity with the university was ripened in a twenty-year professorship in its faculties of agriculture and arts and sciences, and whose acquaintance with the Indonesian language and familiarity with the Indonesian situation were provided neither by UK nor by AID but the Agricultural Development Council. He was a visiting professor of rural sociology at IPB for six years, first as an associate on the ADC staff, next as the chief of party for Kenteam. The book is thus a product of the Beers experience as well as a report of formal study, and its interpretations are the author's temperate conclusions.

A. D. ALBRIGHT
Vice President, Institutional Planning
University of Kentucky

An American Experience in Indonesia

1. Bogor, the Agricultural University, and the Kentucky Contract Team

Bogor, Indonesia
Hospital Street—Djalan Rumah Sakit II—in Bogor lies between the main building of the Agricultural University—IPB—and a row of nine faculty houses. The main building is a place of offices and laboratories built by the Indonesian government after independence of Dutch rule was proclaimed in 1945 and confirmed by international agreements in 1949. The nine houses, and three more a few blocks away, were built for the Americans of the Kentucky Contract Team who worked in the Agricultural University from late 1957 until early 1966.

At the horizon to the left of the Kentucky team houses, a rightward sweep of the eye catches Pangrano and Gede, the two mountains in Java's volcanic midrib that shoulder a pass for the road to Bandung, the southern coast, and the central and eastern provinces. Moving past the university to Mount Salak in the near distance, the eye stops at the big trees which surround and overhang the world-famous biological laboratories and tropical collections of the Great Garden (Kebun Raja). Beyond them stands the palace where Sukarno succeeded the Dutch governors-general in residence. Commercial Bogor, serving the 150,000 people of the city and its village fringes, spreads away from the far side of the garden. The nearest rice terraces are over the hill at the left end of Hospital Street, not seen from the Kentucky team houses, and farther out is the wet mosaic of padi-plots (rice fields) between the mountains and the sea where sixty-five kilometers northward lies the capital city, Djakarta.

In the yards of the faculty houses are coconut, papaya, banana, bamboo, hibiscus, and Christmas roses all year long. In benignly constant equatorial warmth, and watered by the

3

rain of more monsoonal thunderstorms than any other spot in the world, the stage of nature in Indonesia is overset with the props of tropical luxuriance. The quick-changing blends of light, color, temperature, humidity, personnel, and social milieus make excitement and satiety almost monotonous to newcomers who have not yet learned to meet them in slow rhythm. On the faces of Indonesian friends and colleagues are shy smiles of courtesy, respect, and quick-fused warmth. Their response to hurt is merely to disappear. In money and in the fruits of modern science and technology this is the land of the have-nots. In building materials for some kind of a kingdom of heaven on earth, maybe it is the land of everything.

Bogor lies in the westernmost of Java's three provinces, the home of the Sundanese, whose ethnic character and history distinguish them from the Javanese in Central and East Java. The students and staff at IPB are mostly Sundanese, but Indonesia is an archipelago of 3,000 equatorial islands, Java being one of the five largest, and many of the other main groups and areas are represented also: the Atjehnese, the Batak, and the Minangkabau from Sumatra; the Maccassarese, Menadonese, and Buginese from Sulawesi; the Dayak of Kalimantan; the Balinese, the Ambonese, and the Madurese.

Much of the fascination of Indonesia and of Bogor lies in this contemporary diversity and in the culture of successive Hindu, Moslem, and colonial-European inheritances with animistic underlay. The aboriginal population are said to have been Austronesians who came from Southeast Asia in the megalithic period or perhaps earlier. The legacies of all historical periods persist together and are seen in kaleidoscopic variety as revolution—and development—change their mix. Features of modern culture which have continuous lineage from the aboriginal period before 100 A.D. include recourse to the worlds of spirits, the drama of wayang (now performed by shadow play, puppets, or human actors), the music of the gamelan (mainly with gongs and other percussion instruments), the culture of rice, and the care of cattle. These are in the cultural substructure.

In the next general cultural layer, coexisting with and not

entirely supplanting the first, are trade, commerce, Aryan organization, the growing of sugar cane and mango, Sanskrit, Indian arts and philosophy, and the great epics of struggle between the forces of good and evil in the Mahabharata and the Ramayana. All these were brought to Indonesia during the three Hindu empires (Sriwidjaja in Sumatra, Mataram, and finally Madjapahit in Java).

The third major cultural stratum—Moslem—dates from about 1400 A.D. when Gujarat traders replaced the last Hindu empire, subdued the sultanates, and brought them to the worship of Allah—acknowledged now by 90 percent of all Indonesians.

The bearers of European influence, a fourth major stratum of culture, were Dutch trade and government, managed by the East India Company in the seventeenth and eighteenth centuries and by the Netherlands government in the nineteenth and early twentieth centuries. Dutch rule was interrupted by three years of Japanese occupation during World War II, too brief a period to impose another layer of culture but one from which many Indonesians say they learned much—especially the practice of guerrilla warfare and the fact that "little brown people can vanquish big white ones." Just after the war Indonesians fought the Dutch and won their independence.

The University of Kentucky Contract Team

After the Dutch period, after the Japanese occupation, and after the beginning of independence came the Americans, not as invaders or colonial masters or an occupying power, but to assist in recovery and development. It was, of course, in the political interest of the United States that Indonesia's venture succeed and that she become a nation of independent strength which, from the American viewpoint, would not fall behind any curtains of communism, iron or bamboo. Furthermore, economic aid was emerging as a new and major instrument of American diplomacy. After World War II, the Marshall Plan and the Truman Point Four Program (both in 1947) sponsored the recovery of western Europe and instituted technical assistance for Greece and Turkey. The

5

European Cooperation Administration, which administered Point Four, was broadened to form the International Cooperation Administration, and this was the agency which first brought American technical assistance to Indonesia.

Hence, when the University of Kentucky team went to Bogor in 1957, the whole foreign aid movement of the United States as it is now known was hardly more than a decade old. But it had begun to make use of universities as major resources for technical assistance because of their long participation in the hurly-burly of American development and their extensive experience in service to their communities.

The first major university contract for work in Indonesia brought the University of California to Djakarta to rebuild and develop the medical faculty there. This contract is remembered as being eminently successful. The University of Kentucky worked under two later contracts, one to develop the ITB (Institut Teknologi Bandung) and the other to develop IPB (Institut Pertanian Bogor—the Agricultural University).[1] By 1964 there were or had been 151 university contracts in the American foreign aid program around the world, but only 16 percent of them lasted as long as or longer than the Kenteam affiliation at Bogor, and only three involved as large a fund of dollars.

Kenteam was in Bogor under a pattern of arrangements defined by the standard AID-university contract of that period. In this instance the objective was to develop a complete college of agriculture to a level of capability for self-regeneration and growth. The basic agreements were between the members of a "technical-assistance complex,"[2] the government of Indonesia, and the government of the United States, the University of Kentucky, and IPB (at that time a pair of colleges in the University of Indonesia). The agents of the Republic of Indonesia were the ministries involved, especially the Ministry of Education through its Department of Higher Education and Science. The agents of the U.S. government

[1] Other universities engaged in technical assistance in Indonesia were Wisconsin, Indiana, and the University of California, Los Angeles.

[2] CIC-AID Rural Development Research Project, *Building Institutions to Serve Agriculture* (Lafayette, Indiana: Purdue University, Committee on Institutional Cooperation, 1968).

were the Washington and Djakarta establishments of the International Cooperation Administration (ICA), later renamed the Agency for International Development (AID). The project was managed at the University of Kentucky by the Kentucky Research Foundation until 1963, when an Office of Overseas Programs (UKOOP) was established. By another change in 1966, UKOOP became a part of the university's new Center for Developmental Change.

Members of Kenteam, Bogor, were from the University of Kentucky staff—either regular faculty or persons engaged especially for service in Indonesia. They numbered from eight to sixteen at a time and totalled forty-seven during Kenteam's nine-year period of service. In Bogor they were guest professors who worked with their host colleagues in the development of the host institution. Studying the composition of Kenteam is like studying a moving average; there were changes every year as priorities shifted among fields of work. Hence, members of the first team taught the basic and some applied sciences: physics, chemistry, biochemistry, botany, genetics, animal nutrition, and parasitology, with one member to introduce home economics and another to introduce extension education. The members of Kenteam at the time of withdrawal in 1966 had been working almost entirely in areas of applied agricultural science: agronomy (crops), forestry, plant physiology, rural sociology, biology, and marine fisheries. During the years intervening between the first and last teams, the composition included additionally farm management, agricultural engineering, soils, horticulture, animal husbandry, animal physiology, veterinary gynecology, veterinary anatomy, poultry husbandry, food technology, inland fisheries, marketing, agricultural processing, agricultural administration, and veterinary public health. The charge to each team member, under the contract, was to assist in developing the work in his own field and the strength and competence of the whole institution to produce Indonesian agricultural scientists.

The Kenteam project was one of the largest of all university-AID contract undertakings in technical assistance in terms of objectives, commitment of personnel, and costs in time,

7

national currency (rupiah), and dollars. As a case of international interuniversity affiliation, it invites observation and evaluation in a search for insights that bear generally on procedures for international cooperation in development, especially in higher education in the agricultural sciences.

The qualifications of the University of Kentucky for the work at Bogor did not come from extensive prior experience abroad and it might have been argued that the institution was therefore ill-suited to the task. Questions might have been raised also about the alleged ethnocentrism of Kentuckians; in some areas of the state, the word foreigner could mean a native of an adjacent county! Furthermore, it might have been noted that University of Kentucky agriculturists were not specialists in tropical problems. But most other universities in the United States had similar and other limitations, and the American commitment of university resources to technical assistance in developing countries was greater than could be fulfilled by the handful of institutions with foreign experience. Land-grant universities especially had a compensatory qualification in their problem-solving focus and their record of service in American social and economic development.

Among the land-grant institutions were some which were notably competent for work among people with limited resources and with problems of chronic underdevelopment. Herein lay Kentucky's special qualification. Research in the agricultural experiment station for decades had included projects to serve the development of low-income agriculture and subsistence living, especially in the eastern end of the state—in the Cumberland plateau and adjacent Appalachian areas. There were no volcanoes in eastern Kentucky, as in Java, but there were multitudes of people with low income and high family solidarity; Kentucky specialists had worked with rural people in the area of labor-intensive agriculture and its alternatives.

In fact, the challenge of developmental needs within Kentucky had led the university to select, through the years, a staff with motivation for and skill in development. For introductory work in the basic agricultural sciences—an early need at Bogor—Kentucky personnel were as adequate as any to the task. In applied agricultural sciences, Kentucky personnel had not, of course, specialized in rice, sugar cane, coconut, or cloves, in the analysis of volcanic residues, in the control of specific tropical diseases and pests, or in the feeding of livestock from equatorial flora. But they were familiar with the nature and importance of what is often called "adaptive research" and they were skilled in the application of research findings to agricultural production and in the enhancement of rural welfare. The intimate association of extension services with research and teaching gave land-grant college workers a habitual multipurpose motivation to apply their knowledge. It gave them service-rendering skills. Many well-considered plans and policies to stimulate economic development, improve facilities for health, education, and welfare, and provide guidance for rural-urban migration had emanated from the University of Kentucky. The sense of mission from which these things had come may have been as notable as the substantive contributions, and this was precisely a qualification by which Kentucky could be a candidate for service at some spot like Bogor.

IPB, the Host Institution

Indonesia's attention to education was large-scale and earnest from the moment of the declaration of independence and the adoption of a basic constitution in 1945. Illiteracy, then almost universal, was reduced with dramatic speed by ingenious programs of the national Division of Community (social) Education. The spread of elementary schooling was also rapid. Attention to secondary schooling and higher education was more costly and developed more slowly, but was increasingly prominent in national objectives.

The first college of agriculture on Indonesian soil had opened in Djakarta in 1940, only seventeen years before Kenteam went to Bogor, but was hardly more than organized before the Japanese occupation closed it. It was reopened in the first year of independence, conducted by the Dutch faculty, and was moved to Bogor in 1948. A second faculty of agriculture was founded in 1946 at Klaten and Djogjakarta.

9

Formed in the turmoil of battle, it became the University of Gadjah Mada, named for an early hero of the struggle for independence. Until 1959, only these two agricultural faculties existed in Indonesia. In that year four more were established by announcement and directive, but they developed slowly.

When the first nine members of Kenteam arrived in the latter months of 1957, they became guest professors (except for an administrative officer and secretary) in the faculty of agriculture and the faculty of veterinary medicine. These Bogor colleges were then parts of the University of Indonesia, of which the other colleges were in Djakarta. (A college in Bandung had just been separated from the University of Indonesia, and established as an Institute of Technology, where another Kentucky contract team was being installed.) An Indonesian veterinarian educated in the Netherlands, Dr. Titus, was the dean of the college of veterinary medicine. The first Indonesian dean of the college of agriculture was Dr. Tojib Hadiwidjaja, a plant pathologist; previous deans had been Dutch. Dr. Tojib had been in the United States in 1957 as an Eisenhower Fellow to study agricultural colleges and state universities, had conferred personally with ICA and University of Kentucky officials about the terms of the affiliation, and had taken part in selecting Kenteam members. He served as dean from 1957 to 1961. It was of interest in later years to hear Professor Tojib recall his experience in presiding over a faculty in 1957–1958 of mainly Dutch and American professors, the former still a majority. A year later the Dutch were a minority. By a third year the Dutch were gone and there were as many Indonesians as Americans. At last by the fourth year the Indonesians were a majority.

In establishing the colleges of agriculture and veterinary medicine, the Dutch had followed the model of their faculties and institutes at Wageningen. At no time was it thought that they would teach farmers, as American professors had done in their own first colleges of agriculture. Most students were under government-stipend appointment; almost none of them came from village life or agricultural experience. The purpose in Bogor would be to train scientists and officials for other institutions and bureaus. Nor were research or public service to be a responsibility of the faculties. Research was the obligation of special tropical institutes, of which there were twenty-six in Bogor, started by the Dutch in 1870 and lodged in the Ministry of Agriculture. Inspection, control, and service functions in agriculture were the responsibility of special bureaus, also attached to the Ministry of Agriculture. Agricultural and veterinary knowledge and services were expected to serve mainly the state production of export crops, not the population of small farmers (pertanian rakjat, people's agriculture).

The deepening resolve to establish higher education throughout Indonesia was expressed in April 1961 by the formation of a ministerial department of higher education and science. A few months later (December) the nation's first Basic Law on University Education was signed by President Sukarno after adoption by the national parliamentary council. These actions provided a philosophy of higher education and a plan for a national system of universities, with one in each province to fulfill the tridharma—the threefold duty of instruction, research, and public service already acknowledged by the faculties at Bogor.

Within four months, Dr. Tojib, the dean of agriculture at Bogor, became the nation's second minister of higher education and science and took up vigorously the execution of the parliament's intentions. He held this post for two years before communist opposition forced him from the cabinet. Sent to Brussels as Indonesia's ambassador, he was recalled in 1966 to become rector (president) at IPB, but this was after Kenteam had been withdrawn.[3]

Shortly after Tojib's appointment to the ministry and Prof. Dr. Ir. Bachtiar Rifai's succession to the deanship at Bogor, the minister proposed removing the colleges of agriculture and veterinary medicine from the University of Indonesia and elevating certain larger departments and groupings of departments to the status of faculties in a newly formed Institute of Agricultural Sciences. The step could not be

[3] Subsequently, in 1967, Tojib was made minister of estates; in 1968, minister of agriculture.

11

taken at once, however, because there would have been resistance to the establishment of another national university in West Java until each province had at least one. Finally, in September 1963, it could be boasted that each province had at least one national university so Minister Tojib ordered the separation of the colleges of agriculture and veterinary medicine from the University of Indonesia and the formation of the autonomous Institut Pertanian Bogor, a step which resolved important "identity problems." Dr. Rifai was invested as the first rector of IPB on February 12, 1964.

In organization and autonomy, IPB was thus comparable with a university, and the translation "agricultural university" is an accurate rendering of "Institut Pertanian." The difference lay in the technical, applied, and professional nature of an institute's program as distinct from a broader coverage of the universe of knowledge. If the rector's post was vacant, a presidium with a chairman and four others would be appointed by the minister. Upon election and appointment of a rector, associates would be chosen, one each for academic affairs, administration, and community service, and perhaps a fourth for student liaison. In some of the colleges deans likewise had associates with a comparable division of responsibility.

In the process of organization after Kenteam arrived in 1957, sections and departments were formed by splitting, by organizing anew, or by combining, and their number increased as personnel became available to man them. Kenteam members looked even in the first years for ways of grouping departments, but there were more decisions to establish new departments than to integrate old ones. In the beginning, there could be only a few: some were "shadow" departments with interim Kenteam heads pending the return of staff from foreign study. Five plant science departments were made into a botany department when a participant returned who was qualified to serve as head. In the veterinary college, virology and bacteriology were split out from a former department of infectious diseases when heads became available. Forestry got its first head in 1960. Several new animal science departments were formed in 1963, again with returnees to become their heads. The concept of a department with one senior person and a number of subordinates (a European heritage) yielded slowly to the idea of a department with several capable senior colleagues working together and with junior members. Professors, in their insecurity, too often viewed peers as threatening competitors rather than as supportive colleagues. There was a zealous tenacity about the way a professor held any competitive advantage he might have, such as the headship of a separate department.

Methods of allocating budget and paying professors tended to inhibit some change. Five departments could probably get more money than one, so there was some pressure to organize more departments than it would be wise to form. Also, there were increments of pay to professors for each additional course taught and each examination given. These systems of allocation and pay discouraged combining and integrating and encouraged overspecialization of courses and departments.

More staff, more courses, more departments, eventually more faculties—all were expressions of the same expansion. Each was a function of the others and they were all parts of IPB's effort to organize itself. Problems of organization were decided by much faculty discussion and effort to achieve consensus. Curricula were developed, divisions of study were recognized, and eventually colleges were organized. Deans were elected from the teaching staff for two-year periods and presided over senates of senior staff. Numerous functional committees were formed: the agricultural dean's report in 1960 listed twenty-five committees. After the formation of IPB in 1963, a complete system of organization and procedure was arranged, providing the university for the first time with formally adopted governing rules and regulations. These were being revised and readied for readoption in 1965.

Problems of personnel at IPB were met by employing graduates as soon as their studies were ended, by engaging the part-time services of staff members of the neighboring research institutes, and especially by utilizing the members of Kenteam and assigning graduates to carry out advanced study in the U.S. in what was known as the "participant" program. In the later years IPB began to develop a home program to upgrade her own personnel for advanced degrees. During this period also there were staff members contributed by the UN, Australian Volunteers, and the U.S. Agricultural Development Council.

The economic deficiencies at IPB were a visible and constant factor and were even more disturbing to Kenteam than the political instability discussed below. The staff at IPB could receive only such small salaries that none could subsist from that source alone. To be a staff member at IPB was, for most, to be a part-timer, and this was almost as true toward the end of the affiliation as in the beginning. This was a problem of the Indonesian economy, which encountered mounting difficulties throughout the Kenteam period, and its solution in the long run could lie only in general economic improvement. It always seemed to Kenteam that if the problem of inadequate salaries could be solved many of the other problems would disappear. Kenteam reports and letters and the minutes of Kenteam meetings harp on lack of local (operating) funds, lack of any funds for research, and decimation of teaching and research flocks and herds because of feed shortages. At one point in the mid-1960s, Kenteam, with the help of other Americans, supplied food supplements for students, getting their supplies from American "surplus commodities" administered through Church World Service. Two features of this were hushed to forestall repercussion— that students were hungry enough to need the food and that the help was from American sources. The story was told of a communist student leader who was chided for going through the line for the extra food and who retorted simply, "Well, I'm hungry!"

A veteran of Kenteam service, revisiting Bogor briefly in June 1968, wrote to friends:

My first and most obvious impression in the Indonesian visit was the emotional release from the fear, political pressure, and insecurity that were present during our last two years there [1964–1965]. There had been general improvement in roads, in services (electricity went off only once in Bogor when a tree blew down in a big storm and snapped a wire), and in supply of consumer goods (except food). There was a general atmosphere of hope during a time of great maladjustment of costs and salaries, population and food supply, ability to work and opportunity to work. One friend said—"My salary is rupiah 4,600 a month and my transportation to and from my job is rupiah 4,500 per month! How do I maintain my big family on the difference of rupiah 100? My wife's family rice fields, some odd jobs, etc., help." But he laughed and said, "We Indonesians seem to keep going." Another former colleague said, "My salary is rupiah 4,000, and my chicken business gives me rupiah 12,000, but to live, keep children in school (which is not free as in your country), pay utilities, etc., we need rupiah 30,000."

So after all the inputs of the Kentucky affiliation at Bogor, the problem of compensation for IPB staff members was as acute as it had been initially, and hardly anyone could give his full attention and energy to what should have been his only job.

Problems of money were attacked by conventional recourse to the national treasury as allocated through the Ministry of Higher Education. IPB tried to supplement its income by agricultural production (rubber and rice) at Darmaga, and (following the example of the University of Kentucky) by establishing a Research Foundation, which, however, received only limited funding. The time came in 1965 when each faculty, under pressure to escape underfunding, sought independent sources of support. For example, the food technology staff planned to set up some kind of processing plant as a source of income. The Kenteam operation under the AID contract was a major source of support from 1957 until 1966. Equipment and facilities (not including buildings or operating expenses) were mainly purchased with Kenteam dollars throughout the period. The contract budget, which was not used for regular current operating costs, provided $5.4 million for the whole project from August 1, 1957, through January 1, 1967, a period of nine years and five months. The participant program's share was $1.9 million. The budget for purchase of equipment was $719 thousand.

Visiting professors from the staff of the Agricultural

Development Council, Inc., were at work in Bogor from 1959 to 1965; ADC allocations were made for certain research programs in agricultural economics, and for support of field and training programs in extension activities.

The welfare problems of staff and employees were met in various ways, perhaps the most prominent being through payment in kind, or fringe benefits: the distribution of food (mainly rice), fuel (kerosene oil), and material (cloth for garments), not only at the annual time of the Lebaran rituals but on a monthly basis, as wages in-kind. Housing for staff, also given as wages in-kind, was difficult to provide and became more scarce as returning participants qualified for occupancy. New houses at Darmaga and in one other neighborhood complex afforded some relief. To Kenteam, of course, the provision of food, fabrics, fuel, and housing as compensation in addition to wages was a new pattern of relationships between a professor and a university.

The issue of internal versus external organizations of students was met by IPB with the view that only bona fide IPB students, as organized in a representative student council, could legitimately be heard on matters of university program and policy. The external communitywide and national organizations were a feature of Indonesian development, however, and they pressed for influence at IPB also, with success in the politically active months of 1965. Shortly thereafter, students —the generation of '66—assumed their new role as powerful political activists throughout Indonesia.

Although it was forbidden for government employees, including staff and students, to belong to political parties, the major parties informally sponsored student organizations which were known to be politically related. The influence of these groups rose and fell with the regional and national ascendance or decline of the parties concerned. In the last few months of Kenteam's presence, the staff and student bodies of IPB were wracked by the powerful struggle of communist-related groups and opposing bodies, reflecting the larger contests throughout the nation, in which anticommunism would shortly prevail.

IPB staff members often asked Kenteam members about

student roles in American universities. They were interested to know about how students organize and for what purposes, how athletics are organized for students, what facilities such as student centers and eating places are provided, and about ROTC training. The return of participants from U.S. study brought concrete information on these points. Competitive sports—especially soccer—began to develop. A cafeteria was established and military training was instituted. It is ironic that these questions were directed to Kenteam in a period just before student activism became prominent at American universities.

The changing political situation in 1965 brought about changes at IPB. A new rector led his colleagues toward independent Indonesian action. He informed Kenteam members that they would no longer attend Senate meetings, which would henceforth be conducted only in the Indonesian language. Thus, he began to push almost imperceptibly toward eventual American withdrawal, scheduled for the following year. The spirit of self-sufficiency, standing on one's own feet, expressed in the acronym "berdikari" (berdiri diatas kaki sendiri) was the prevailing mood of the national leadership, antagonistic toward western culture and "Nekolim" (neocolonialism, colonialism, and imperialism). This orientation was expressed forcefully in President Sukarno's declaration of withdrawal from the United Nations. The language of the commitment to self-reliance, however, expressed a well-established American value, that of respect for any effort at independence and self-support.

One of the major reorganizations of IPB occurred during the last months of Kenteam's presence in Bogor, quite without any direct American participation. Departments, reorganized to number about twenty-three, were removed from the six faculties and made directly responsible to the rector. The faculties, each with a dean, were to have responsibility for curricula, instructional programs, and the students enrolled therein. The departments were to provide teachers as needed by the faculties. The departments also were to have responsibility for research and for public service (extension) activities. Kenteam saw in this an interesting new arrangement blending some of its own preferences into new forms, and this is the way ɪᴘʙ was being organized when Kenteam left Bogor.

Other features of organization were the beginning of an association of alumni and the establishment of a Board of Curators to perform a trustee function, linking the institute with the community, the province, and to some extent even the nation, for ɪᴘʙ was becoming a national center of higher education in agriculture. This Dewan Curator, an appointed public board, was first established in 1964 to serve more in public relations and in advisory roles than in policy making.

"The Situation," 1957–1966

Kenteam's first members moved into their Bogor houses in late 1957, only twelve years after the Sukarno-Hatta proclamation of independence, only eight years after sovereignty passed from the Dutch. The Republic of Indonesia had existed by that name only since 1950, which was the year also of Indonesia's admission to the United Nations. It was too early for stability; uncertainties were characteristic of all the Kenteam years in Indonesia, and these were sensed as a fluctuating background of complications which inhibited adjustment and retarded accomplishment. The initial cordiality toward Americans was partly a rebound—a compensation for the rejection of the Dutch upon whom there had been dependence before. This was inevitably corrected by the growth of Indonesian self-assurance in nationalism, by the continuing practice of independence, and, on another level, by the increasing power of communism behind the National Front. There was the general truth, of course, that Americans, like the Dutch, are orang belanda (white men). Eventually the affection of Indonesians for Dutchmen was reasserted, just as were the Indian affection for Englishmen and the Indochinese affection for Frenchmen. The love-and-rejection ambivalence of those newly free for their former masters could not apply to the Americans for there had been no previous subjugation by them. During the windu (an eight-year period) of Kenteam's life at Bogor, however, the politicians in the left corner were moving out toward the center, and they mobilized a new sentiment against Americans, lumped with the British and others under the term Nekolim. In general and most of the time, ipb was an island of acceptance, mainly cordial and appreciative, and Kenteam could live nonpolitically and work professionally at Bogor where nationalistic tides of hostility to Nekolim subsided or were diverted until the last months of Kenteam in 1965 and early 1966.

In conversations, especially between Indonesians and Americans, the climate of uncertainties and political turbulence was generally mentioned as "the situation," a neutral phrase which all could use without implying either opposition or endorsement. But in "the situation" Americans confronted elements of resistance and threat which contradicted the normally cordial and supportive relationships in the university and among friends in Bogor. "The situation" identified itself to Kenteam by an intermittent succession of crises and disturbing events which began in the early years and continued throughout the middle and later periods of the affiliation with ipb. There was hardly any time when Kenteam families could feel completely relaxed in the charged atmosphere of contemporary Indonesian history, variously antagonistic to western nations, including the United States.

In the year of Kenteam's arrival, Sukarno intensified his campaign to wrest control of Irian Bharat (West New Guinea) from the Dutch. The United Nations declined the role of negotiator and Indonesia seized many of the remaining Dutch interests. Tension over this issue was continuous for the next five years until Irian Bharat was won. Kenteam members were apprehensive, and the newly arrived first chief of party in Bogor wrote to Lexington in November 1957, "While the campaign seems to be directed primarily against the 'Dutch' the man on the street sometimes thinks of all 'blonds' as Dutch. Mistaken identity may result in unexpected incidents and, therefore, our caution against unnecessary circulation outside of our neighborhood, especially after dark."

A declared state of war and siege enveloped Bogor in that year and this restricted internal travel by which Kenteam members might have advanced their orientation to Indonesia and its tropical agriculture. It was in this first period (1958) also that the Permesta rebellion against the Sukarno government broke out in Sumatra, Sulawesi, and parts of Java. There was extensive fighting, rebels were pushed into the hills, curfews were imposed, and for the next three years travel to and from Bogor was further curtailed. By the end of this first period of the Kentucky contract (1959), withdrawal of the Dutch staff members from the faculties was completed and Sukarno had assumed the relatively unrestricted powers of a president-premier. Kenteam members were vexed by interferences with mail—censorship, and some nondelivery—startled by a peremptory devaluation of banknotes, and puzzled to understand the forced movement of Chinese businesses from rural areas to urban centers.

The Kentucky-AID contract was extended beyond the initial three-year term, and it was at the beginning of the next and middle period of the team's tenure that Indonesia's elected parliament was dissolved and replaced by an appointed Peoples Provisional Council (1960). Especially disconcerting to Kenteam Americans was the dissolution of organizations such as boy and girl scouts and Rotary clubs because of alleged foreign domination.

During these middle years a national eight-year plan for development was announced (1961) and all programs and projects were expected to orient themselves to this collection of schemes, which was more a product of political discussion than of economic study and technical assessment. In this period also there was an upsurge of anti-American sentiment after the 1961 death of Congolese Patrice Lumumba, to whom some Indonesian leaders accorded the status of hero in the struggle against colonialism and imperialism. At such times American study for Indonesian students was criticized by the newspapers and by public figures. Demands were made by political leaders that no more students be sent and that those already in the United States be recalled. (Actually, more were sent each year until the last—1965—and there was no recall until late 1965 and early 1966.) Anti-American sentiment was fanned again in these middle and later years by alleged United States persecution of Black Muslims, by U.S. political and military action in Laos and Vietnam, and by U.S. reaction to the India-Pakistan war.

Early in 1962, the political temperature rose again in the Irian Bharat campaign and in February a Djakarta mob broke windows in the U.S. embassy. There was a subsidence by May, when Irian Bharat was actually turned over to Indonesia and it was known that U.S. mediation was a major factor in getting the matter settled. The feelings against America went up again, however, and reached a momentary fever pitch in October because of the Cuban missile crisis. During that incident Kenteam was alert to the possibility of evacuation.

A new crisis at the beginning of Kenteam's final three-year period reached Bogor in May 1963 in the form of anti-Chinese rioting in which students, including many from IPB, were the activists—their first participation in such violence. The first rioting took place in Tjirebon and Bandung, occurring not long after a highly dramatized visit to Indonesia by the president of mainland China. Kenteam members stayed in their homes and were themselves under no threat, although they were perplexed and upset by this turn of affairs which resulted in destruction of property, especially Chinese shops, although it did not include attacks on persons. Explanations given to Kenteam varied. One theory attributed the action to opponents of the government who aimed to discredit it by the embarrassment of indirect attack which dramatically protested poverty and inflation. Other theories attributed the action only to student unrest, or to racism as such, or to resentment over the economic advantage of the Chinese community. Kenteam members who had Chinese friends were emotionally caught between loyalties. Additionally, and from this time more or less continuously until the end of the affiliation, Kenteam was upset by the disturbance of class and laboratory routines and the diversion of student interest from study.

The frequent diversion of student attention from the main business of university study was a manifold problem. Many students had insufficient nourishment and lacked energy for hard work; costs of clothing, books, and food were beyond the reach of many; they were attracted to political groupings and divided among the organizations that were formed; they were organized for military training under Hansip (Pertahanan Sipil, civil defense) operations. Members of Kenteam, who felt that they had travelled a long way for a short time and at considerable expense, wanted to work intensively with students and were frustrated by these complications. Later, another diversion of student attention was brought by the national campaign to increase rice production, OPSSR (operation self-sufficiency in rice) which resulted in sending students to the provinces to stimulate the production of rice cultivators.

But Kenteam's main trauma in this final period came with the spectacular advent of "Konfrontasi," the Indonesian campaign of confrontation to crush (ganjang) Malaysia, which led, among other things, to the burning of the British embassy in Djakarta in September 1963. It seemed that further political divergence of Indonesian and American views was inevitable, at least in the short run. Representatives of the Indonesian and U.S. governments had just undertaken a joint program to seek economic stabilization, but this was rendered impossible by Konfrontasi. Resources could not be mobilized toward any goal of economic stabilization and the major efforts of the U.S. Agency for International Development became pointless and futile, at least for the time being. Effective contacts between Americans and their Indonesian "opposite numbers" began to dwindle and atrophy. A letter from Kenteam to Lexington reported, however, that "recent flurries of security/insecurity have found the Bogor Team families relatively un-upset; we have been a little apprehensive, of course, but apparently not as worried as some of the American families in Djakarta and Bandung where British and Australian neighbors were evacuated." The campus coordinator of the contract project replied that, "I don't suppose any event in the past several years has so concerned our participants [IPB staff members who were studying in the states] as the Malaysia affair."

Throughout these final years "the situation" kept Kenteam

families in suspense. There was a token book-burning in Djakarta in May 1964; books of Dutch, German, and English publication were destroyed in symbolic repudiation of "imperialism and colonialism."

The minister of higher education, Professor Tojib, had intimated in discussions in January 1964 that the University of Kentucky and AID might be invited to extend in time and expand in scope the work in higher education in the basic sciences, the engineering sciences, and the agricultural sciences. A survey of status and need was projected in which pairs of Kenteam members at Bogor and Bandung and their Indonesian colleagues travelled to the national universities throughout the republic. There were immediate interruptions, however. Less than three weeks after the planning conference with the minister a national food crisis came to the fore, and the minister was preoccupied with food shortages in central Java. The drought was more severe than any since 1924—wells were dry in the middle of the rainy season, villagers were flocking into the cities for food. Some suspected that the PKI (Partai Kommunis Indonesia) got the trucks and pushed this movement of refugees. Others said the lurahs (village heads) were hiring the trucks to haul people out. The minister, having himself received orders from the president and cabinet, instructed the rector of IPB to organize an effort to obtain substitute foods using Indonesian resources and not by the use of foreign aid. The rector wanted the wives of Kentucky team members to help in developing recipes for palatable uses of the foods—which they did—but he did not formally involve team members in the emergency discussions. Their participation was only informal as they were approached by their Indonesian colleagues.

The plan for the survey of status and need was drawn up and most of the travelling was done cautiously and inconspicuously in April and May 1964. But the scheme lost momentum and finally was suspended, not by formal termination but only by inattention. Allegedly too "close to the West," the minister had been under attack from the moment of his appointment in 1962 and was finally removed in 1964, about a year after expanding the two Bogor faculties of the

23

University of Indonesia into the Agricultural University, IPB. Kenteam had taken great interest in the survey of possibilities for a larger task in Indonesia and swallowed the "forced landing" with disappointment.[4] AID had been a skeptical participant in the planning for enlarged and extended activities and was acutely sensitive to U.S. congressional questioning of "what was going on in Indonesia," and Kenteam members sometimes wondered whether they were living on time borrowed from both the forces of Indonesian history and the domestic U.S. jaundice over foreign assistance. The new minister, at a first anniversary ceremony for IPB, announced that the Kenteam contract would not be renewed because "nine years was enough." But a large group of IPB staff members went to the U.S. in May 1964; it was the largest group yet to be sent and proved unexpectedly at that moment to be the last, until after Kenteam itself had departed.

What appeared on the surface to be internal personalized scandal at IPB in the summer of 1964 (and no doubt under-the-surface political struggle was involved) shook Kenteam at least with mild tremors in midsummer. With well-schooled restraint the Americans "kept out of it" completely, as they had learned to do in all the multisided conflicts of national or local scope. In the fall (but summer and fall are American time-concepts; the even course of Indonesian climate does not have these seasons!) school opened as usual and the tempest seemed to have subsided. Kenteam was restive at new harassment in communication—there was a period of irregularity in APO services which normally brought letters from home to Djakarta every Tuesday. Kenteam files have a number of letters from this period in which the administrative officer reported whimsically the arrivals and nonarrivals of "the big blue bird," which frequently did not risk landing in Djakarta when there were threats of sabotage. There continued to be anti-American demonstrations in Djakarta at the embassy, the ambassador's home, and the USIS library, and

[4] American attention was drawn again to the idea of a national program for higher education in agriculture by an AID survey team in Indonesia in January 1968. The report of this team was eventually implemented in an AID contract with the Midwest Universities Consortium for International Assistance (MUCIA).

similarly in other cities (Surabaja, Bandung, Medan). The only newspapers were virulently anti-American in tone and substance, but only two or three were published in English and most Kenteam members did not read Indonesian.

Tension-producing developments accelerated in the final year of residence in Bogor, 1965. One event in January and February was particularly critical for Kenteam, whose chief of party had addressed a memo on Kenteam viewpoints about "Study for Advanced Degrees" (later dubbed the SAD memo!) to the rector and his associates. The memo was soon discovered to include certain evaluative comments which proved unexpectedly provocative. The memo came into the hands of politically active leaders (to the left of Kenteam) who seized the occasion to address a resolution to President Sukarno charging insult to Indonesia and demanding expulsion of Kenteam. Bulletin-board copies of the resolution were observed by Kenteam members and a quick succession of responses was triggered. There was coincidentally an unanticipated transfer of authority at IPB which had its influence on the outcome of this incident. The memo had been prepared January 5 and was delivered to the rector January 11. The rector, following a resurgence of the past summer's turmoil, was removed three days later and assigned to a post in the ministry at Djakarta. The minister installed a presidium of five members to direct the affairs of IPB and to conduct the process of selecting a new rector. Hence, the SAD memo had to be acknowledged by the incoming presidium, not the outgoing rector. Kenteam's chief of party was alerted informally that there would be trouble over the memo. He sent notes to ask that the memo be returned for correction, but only one copy was returned and the "fat was in the fire."

Within a week one Kenteam member was advised by a colleague to abandon plans for a trip to one of the islands, and on that date, January 22, anti-Kenteam resolutions appeared on the bulletin boards. The presidium presented Kenteam with an ultimatum demanding apology by 8 P.M. that day. Kenteam submitted a letter regretting having occasioned the unanticipated and completely unintended hurt which the memo had engendered; the next day the presidium formally accepted the apology and the incident was officially over. But the smouldering aftereffect persisted throughout the rest of Kenteam's presence, and influenced, in ways of which Kenteam could never be sure, the effectiveness of the work. Kenteam—previously successful in avoiding visibility as a political object—had now become one and there was no possibility of reversal. The minister was said to have advised the rector to ignore the pressure for expulsion of Kenteam, "keep the lid on," and avoid discussion of the issue. But it was beginning now to be reported that Bogor faculties were under great threat from political opposition because of their tolerance, among other things. Bogor wives of Kenteam participants in the U.S. were threatened for visiting with Kenteam wives and began to stay away. Sensing this condition, Kenteam began to avoid friends to prevent the "kiss of death" phenomenon. Servants in American homes were chided, even by students and laborers at IPB, and urged to "ganjang" (crush) their employers. (None of them did!) One Indonesian colleague who continued to visit American friends, but only after dark, said, "Now is a time when you must smile at your enemies and don't talk to your friends." A withdrawing technician for the UN had been told that Americans would be kept in Bogor as long as possible, however, to get supplies and equipment from them. It was said by some that they accepted the personnel only in order to get the material.

Within a few weeks after the SAD affair there were new disturbing factors: increased harassment of Americans in Djakarta and other cities, the takeover of the Goodyear Tire Company factory in Bogor, electric and gas meter pullouts at American buildings in Djakarta, a postal union boycott of Americans, and resolutions by a national organization of college graduates (HSI, Himpunan Sardjana Indonesia) to send American lecturers away and recall Indonesians from study in the U.S. It became known that IPB staff were frightened by threats directed at their jobs. In June, the cost of living went up another 50 percent. In August, Sukarno announced the doctrine of the Indonesia-Peking axis.

The growing estrangement between the governments of

Indonesia and the U.S. reached a condition of near-maximum strain short of breaking diplomatic relations in the spring of 1965. The Agency for International Development was quietly but rapidly withdrawn in the last weeks of the 1964–1965 fiscal year, until its residual interests were handled by an AID affairs officer in the embassy. By September 1965, only the two University of Kentucky teams of professors at Bandung and Bogor remained of the earlier projects to serve Indonesian universities through affiliation. Sister agencies also withdrawn included the U.S. Information Service, the Military Technical Assistance Group, and the Peace Corps. Certain private bodies, notably the Ford Foundation, also closed their offices in Indonesia.

In American quarters, the university affiliations, where Americans and Indonesians were still working together, became most-favored projects—which was much more pleasant than being the tolerated contractual step-children of a foreign aid bureaucracy, as had seemingly been their status only a few months earlier. In Indonesian quarters, too, the affiliations had greater survival value than other technical cooperation projects. The heads of ministry and university units could see better than others through the clouds of political discussion to identify the contribution of the affiliations to growth and development in higher education.

The withdrawal of American efforts was preceded and accompanied by the elaboration of President Sukarno's new insistence on Indonesian self-reliance. The first member ever to withdraw from the United Nations (1965), Indonesia embraced the newly named but anciently formulated concept of self-sufficiency, Berdikari—standing on one's own feet.

During the period of the AID withdrawal, there were opposing views of whether the university affiliations should also terminate ahead of schedule. Arrangements for closing the university projects had been nearly completed by AID officials when the contrary judgment of Ambassador Howard P. Jones intervened and prevailed. His view was implemented in May 1965 in a joint memorandum of agreement between Ambassador Ellsworth Bunker, who was in Indonesia for only a few days on a special assignment, and President Sukarno which provided that technical cooperation in the universities would continue to the end of the existing contracts. Only four contracts then remained in force: the medical education affiliation of the University of California with Airlangga University at Surabaja; the engineering contract of UCLA with Gadjah Mada University at Djogjakarta; and the University of Kentucky contracts—one in the engineering sciences at ITB in Bandung and the other in agricultural sciences at IPB.

Not long before these developments, it had been thought in both the related Indonesian and American educational circles that the work of these affiliations might actually continue in some form rather than end, or might be followed by appropriate "second generation" contracts. But political relationships continued to deteriorate and the UC-Airlangga team actually left Indonesia within a few weeks after the Bunker-Sukarno agreement. (The UCLA-Gadjah Mada project was ending in 1965 and its withdrawal was routine.) Another moment of decision came in August during the buildup of pressures which exploded in the attempted coup of September 30, 1965. It was then (in August) arranged by Ambassador Marshall Green (who succeeded Jones on the latter's retirement) and the Ministry of Higher Education that the last two remaining teams—Kentucky at Bandung and Bogor— would withdraw after the first semester of that academic year, rather than on June 30, 1966. Very shortly after this adjustment, it was ordered that dependents of the Kentucky team members should be evacuated. During their final four months in Bandung and Bogor most of the Kenteam members maintained residence without wives and children. Only wives who were still "in the region" while under evacuation orders were allowed to rejoin their husbands in Bandung and Bogor after about six weeks of absence from Indonesia and after the security implications of the postcoup trend had been more fully assessed.

For most of the American and Indonesian educators working together during the affiliations, it was clearly a somewhat reluctant disengagement, required only by political exigency and not by principles of educational development. But Kenteam, withdrawn by the United States government with the agreement of Indonesia and the University of Kentucky yielded to "the situation" and left Bogor and the Agricultural University in March 1966.

The university-contract stratagem in technical assistance had had a nine-year test in a context of economic and political instability and turbulence. Planned initially as a three-year project but inevitably extended—for no college can be built in three years—the work of the affiliated institutions and governments had made two faculties into a five-faculty autonomous university, had provided advanced education for a majority of the staff of teachers, and had organized departments and curricula. The objective records of size and scope indicate that at least the minimum purpose of the affiliation had been accomplished. But how substantial, really, was the achievement of these Americans, who were mostly without previous developmental experience abroad, who had to overcome perplexity about the ways of their colleagues and the norms of Indonesian living, and who had to adapt the skills they had been applying at home and improvise others? The chapters which follow are a true-story report about the persons, principles, and procedures by which professors from temperate Kentucky worked in tropical Indonesia.

2. Kenteam's Relationship with IPB, the Agricultural University at Bogor

The Indonesians were a people new to authority, independence, and responsibility, and their revolt against the Dutch was severe and complete. Their acceptance of technical assistance could hardly be without suppressed chagrin and concealed resentment—not directly against the Americans but deflected toward them from the real object, which was their humiliating need for help. Americans can probably never know or understand completely the Indonesian appraisal of Kenteam and its work, nor can Indonesians understand the depths of Kenteam response to them. Surface and middle-depth relations, at least, can be identified and their impact on IPB's development can be examined. The themes of Kenteam-IPB relationships appear in all sections of this report; certain conspicuous features, however, are brought together here for special review.

IPB Reception of Kenteam

It was a part of the concept of the project that each American would become a member of his department or unit of IPB and that his work would be conducted from that vantage point. Each Kenteam member had an office location in his department, either alone or with Indonesian colleagues. But IPB departments extended several types of reception to Kenteam members. Some assimilated them as co-workers having equal status with other members. Some were more constrained and tended to isolate an American member, limiting him more or less to giving lectures in an agreed course of studies. For some American professors initially there were no departments until Indonesian staff returned from overseas study. In some departments, Indonesian members subordinated themselves to an American member and treated him as some kind of special head. In other departments he was accepted as an adviser in all matters of any concern to them. The departments, in their faculties and within the expanding institution, were as varied in nature as in a university of comparable size and complexity at home.

There was a once-only orientation of Kenteam members, those who were in Bogor in 1962, arranged especially by the rector of IPB. This was a well-presented and much appreciated sequence of lectures, discussions, and local trips on occasional days within the range of a few weeks. Americans are quite insistent on orientation for visitors from other countries; conversely, as guests abroad Americans profit in the understanding and adjustability that orientation by host groups can readily produce. Kenteam members could have been greatly helped by more special orientation from their Indonesian colleagues—to convey understanding of Indonesian history and culture, introductory acquaintance with Indonesian government, with the programs of agencies and groups, and with itineraries of internal travel in Java and some of the other islands.

The agent-client model might be applied to the Kenteam-IPB relationship, but it is not really a good fit. It implies that the agent who "knows something" is thereby superior to the nonknowing client. Also, to think of the IPB-Kenteam relationship as "bilateral" is to structure it with an American side and an Indonesian side, but there was never such a simple bifurcation as this. The clarification of objectives, the organization of staff, the planning of curricula, and the acquisition of physical properties were manifold issues. Differences of understanding and viewpoint are usually present among the members of any staff, and they may become more varied as the institution grows in size. Defenders of status and proponents of change align themselves not only by nationality but by various postures of intellect and culture as well. A model of jointness or mutuality is more appropriate; this establishes a relationship in which the two come up together before a problem in which they have joint interest and which they jointly seek to solve. Each learns and each teaches; each helps and is helped.

Also appropriate is the host-guest model, because each Kenteam member was a guest professor (guru tamu) at IPB, and was so identified on the official rosters. The connotation of host acknowledges the proprietorship of the Indonesians in their own developing institution and the patterns of hospitality and courtesy with which they accepted Kenteam. The concept of guest conveys also the meanings of humility and appreciation that are expressed by a visitor. The host-guest model indicates a donor and recipient relationship that puts the Indonesians in the former and the Americans in the latter role. This prevents or corrects any construction that would identify AID or Kentucky as a donor of assistance.

Communication: Points of Strain and Accommodation

Kenteam members were not required to learn or even study Indonesian, although all were encouraged to try. The objective was less to acquire skill than to give evidence of interest in Indonesia and actually to enlist and strengthen such interest. Members were reimbursed for payment to Indonesians for coaching in language. Tapes and mimeographed materials were provided from AID in Djakarta. Several members thus acquired some ability to communicate with their servants, a few were able to take part in simple conversations and to read a little, but none became fluent enough to think and work in Indonesian. Indonesian colleagues, on the other hand, almost never assigned duties of interpretation to anyone, relying with pride on their own knowledge of English, which they had studied in middle school and college. Many times, since Kenteam knew so little Indonesian and Indonesians were participants in the discussion, without especially assigned interpreters the Americans sat through meetings and conversations with limited—or no—understanding of what was going on. This clumsiness in communication was a weakness in the whole relationship.

While language was often a problem, differences in mother tongue were only part of the total problem of achieving understanding. Also involved were several features of the etiquette of communication. "Going through channels" is essential in all societies, but each culture has its own chains and systems of communication protocol. These reflect, among other things, the characteristic patterns of rank and respect. One illustration of the proper use of channels was the requirement that any external request for the services of a Kenteam member had to come first to the appropriate dean for his sanction. Perhaps no objection would have been registered if this procedure had been skipped, but the recognition of the dean's authority was correct and did not go without appreciation.

In Indonesia, and especially at Bogor, there were the additional confusions of mixed ethnic origin: there were students and staff members from nearly every island and tribe. Dominant traits were common to most of the systems of communication, however, and doing things correctly by Javanese or Sundanese traditions was fairly good insurance against major mistakes in other groups.

Not all American professors are well practiced in orderly analysis and presentation of needs, but all have at least been oriented, if not somewhat habituated, to these procedures in their own institutions. Kenteam was therefore continually baffled by Indonesian customs of inquiry and response, of asking and giving. It came to be said by Kenteam that Indonesians would never make known their wants if there was any chance of a negative response. Americans, on the other hand, would never promise an affirmative response until they knew what was wanted and planned. Between the American and Indonesian groups, therefore, there was always a potential failure in understanding. Someone likened the situation to that of a sparkplug whose gap closed only when the current jumped across. Americans wanted their IPB colleagues to formulate written expressions of need, plans for courses, for departmental growth, for advanced study by young staff members. Indonesians were often content to communicate these matters in conversation and conference, sometimes by spoken inference, leaving them with no problems of face saving if responses were indifferent or negative. If an Indonesian made a request and the answer was not "yes" but a question, he would consider the response negative and would close the matter or bring it up again only in some very different form and context. Also, if an Indonesian received a negative reply or otherwise "lost face," he would not dispute the conclusion but might leave, not to return again to the source or the scene of the rejection. It was necessary for Kenteam to understand this or risk losing effectiveness in associating with IPB colleagues.

Kenteam was often uncertain of what IPB needed, and IPB staff members were uncertain of what Kenteam members would approve. In prolonged association, of course, each learned enough of the other's customs to understand requests and answers fairly well, and to tolerate breaches and compromises. IPB staff sometimes made special requests for help from Kenteam. They wanted training in administration; they wanted a course in scientific writing; they wanted an advanced seminar in research methods; they wanted a paper on the role of an agricultural college in a developing nation. Continuous working contact was maintained, even to some extent during the last tense months before Kenteam's departure. There were members of both groups who came to understand the gap in communications and who worked to bridge it by reciprocal adaptation—the Indonesian speaking more directly and the American becoming more sensitized to indirect expression, allusion, inference, and "cue reading."

What Kenteam members at first considered dishonest they came to recognize as courtesy and consideration. Many a genteel Indonesian, like other Asians, considers it his duty to please a companion, guest, or host, by saying what he thinks the listener wants to hear rather than "the truth," if that be unpleasant. A leading question implies a particular answer so that is clearly the answer to be given, in large part as a sign of respect. Agreement is highly valued!

A notable fact in the years of affiliation between IPB and Kenteam was that relatively few requests came from Indonesians for special favors for themselves or their friends. There was an occasional request for special transportation or for preference in the arrangement of a trip to the U.S., but IPB

34

staff did not seek books or commodities or preferential travel for themselves. Perhaps American tendencies to say "no" were in such contrast to the Indonesian etiquette of indirection that there was general unwillingness to be the recipient of rejection or denial.

The speech of Indonesians, whether in their own language or in English, showed deference to valued persons: friends, guests, superiors. Forms of speech conveyed respect. Slow speech, soft speech, subtle speech, and the use of special words made up a communicative art rather than simply efficient interaction. Speculation suggests possible reasons for this artful guile in Asian speech, a quality which Kenteam had to learn to understand. There has been time to be subtle; no hurry, why be direct? Subtlety is protective; it provides a defense of self and protection against loss of face. Subtlety is courteous; it helps others to maintain their defenses. Subtlety is an art lending grace and beauty to communication, and one may seek skill in artful expression.

The intent of subtlety is sometimes to deceive, but only sometimes. It is part of the art of speech; thus its purpose is to communicate, not to prevent communication. Written language depends heavily on subtlety. Most Westerners write, but few speak, with beauty or eloquence. But the corners and curves of subtlety are understood by those who practice it. Its indirection is direct enough for those who use it. We who are direct are the only ones obstructed, and part of its function indeed is to obstruct us. But whatever the explanation for subtlety, it is now a cultural compulsion and informal discourse has become ritualistic. To violate subtlety is a breach.

The American members of Kenteam and their Indonesian colleagues approached the identification of problems and the search for solutions within different cultural and psychological frameworks of explanation, and neither group could fully understand all of the other's assumptions and hypotheses. Americans, whose habits involved recourse to rationality, logical thinking, controlled experimentation and verification, were only occasionally and vaguely aware that sometimes their Indonesian colleagues gave credence to instruction from the spirit world, conveyed by the dukun (shaman) through the exercise of his special powers of communication. Rational behavior was only one of several modes of approach acceptable to many of their associates and neighbors in Bogor. Some Americans listened dubiously—but respectfully—to stories believed by students and staff members of what was brought to pass by elves and ghosts. They observed an occasional name-changing ceremony which was intended to, and probably did, convert some insoluble problem to a simpler one. A few Americans were appreciative guests at Indonesian "selametan" ceremonies of communion, propitiation, and thanksgiving. They saw the heads of bulls buried according to ritual under the thresholds of new buildings. They accepted dates for ceremonies that astrologers judged to be suitable. They learned that it would not be wise to live in a house facing east. They heard explanations of erratic behavior which could be forgiven because it had been caused by black magic for which a victim need not be held responsible. They knew of changes in fortune which occurred following ritualistic appeasement of the spirits of the former occupants of the roooms, houses, dormitories, or offices. For any unexplained event or problem, many Indonesians had recourse to more alternative explanations than an American professor could consult. (If development is defined as increasing the number of alternatives, who is the most developed?) These differences in systems of causation, although not usually known to be operating in conversations and conferences, were nevertheless occasional barriers to complete understanding between Kenteam members and IPB colleagues. No doubt, the two groups were sometimes like space ships which sought to dock but were not even in the same orbit. No member of Kenteam ever penetrated more than superficially into the rich mysteries of Indonesian explanation. Probably no adult could be an initiate in these systems without having been an Indonesian child. The structure of scientific thought, in cognitive dissonance, could probably best be added to the other frameworks of explanation as an additional, not a substitute, intellectual resource.

Strange to Kenteam was the custom of decision-making by

consensus rather than majority vote. This seemed to be the result of diffusion of responsibility so that no spokesman could be found. Had Kenteam recalled the practice of the Quaker meeting, in which the sense of the meeting is voiced by the clerk, they would have found only a partial analogy in their own culture. But it seemed to Kenteam members that decision-making by consensus is not as appropriate in a day of science and technology as it may have been before modernization. Agreement in opinion may be appropriate for issues that can be decided on the basis of opinion. But what about knowledge? Where the facts are common knowledge, each member of a group may be as capable as every other of participating in decision-making. But where the facts are special and not known in common, decision by consensus may yield conclusions which are unrealistic. To Kenteam, decision by consensus seemed really to be decision by a dominant leader and acquiescence by others in deference to the leader's rank or power or superior knowledge.

Rumor was another problem in communication faced by Kenteam. In a developing society which has had to rely on informal, incomplete, and delayed messages, rumors, more often wrong than right, are widely prevalent and more influential than in societies with modern technologies of communication. Village and tribal people are marvelously facile with grapevine, drumbeat, and runner, but the content of their news is personal, primary, and immediate. With modernization and with better machinery for communication there comes also a new complexity of messages until there is a surfeit instead of a shortage. Perhaps someone should study whether a surfeit of news brings a society back to rumor as a false mode of simplification. In any case, in Bogor Kenteam lived in a bewilderment of rumor. The subjects were often predictions of economic and political action, of shifts to left or right, of changes at IPB, and the usual reports of personal virtue or alleged misconduct. Probably no Kenteam member kept a record of rumors to check later their fulfillment or disappearance, but surely the falsity rate of rumors was very high. It behooved the Kenteam member to keep his poise within swirling rumor, whether from American or Indonesian sources, and to hesitate as long as need be for verification and solid information. Misjudgment of rumor or overhasty response could be damaging to the harmony and understanding of relationships, both among Americans and cross-culturally, and this imposed on Kenteam members extra obligations to be patient, obligations which were usually, but not always, acknowledged and accepted.

On Keeping Informed

There were times when Kenteam members felt somewhat lost and ineffective because they didn't know what was "going on." They expressed to each other and probably to others a wish that Indonesian colleagues would keep them better informed of plans and events. Before the merger of colleges to form IPB in 1963, the deans and Kenteam's chief of party had frequent but not routinely scheduled communication. Each separately initiated contact according to inclination and each kept his colleagues generally informed of matters discussed. After 1963, it was the rector (or the chairman of the presidium in periods when the rector's position was vacant) who took deliberate steps to inform Kenteam of developments and situations when he considered it appropriate to do so. Whoever held the office would call for—or upon—the chief of party, not frequently, but occasionally, to speak of meetings with the minister, to swap interpretations of political events in the U.S. and Indonesia and to discuss their implications for Americans in Bogor, to explain faculty views and decisions, to inquire about Kenteam interests and morale, to engage in discussion of some idea or to solicit reaction to some proposal under consideration and to "drop cues."

In the earliest days university senate meetings included nearly everyone; later only the section leaders and department heads attended, eventually mainly the deans. Formal faculty meetings were infrequent; informal sessions of ad hoc groups of the Indonesian staff occurred more frequently. In the first years of the affiliation, Kenteam members were prominent in attendance and influential in participation. They were rapidly outnumbered as the faculty expanded and their participation was less prominent as faculty organization progressed, until at the end the members of Kenteam were operating mainly within their departments, and the meetings of faculties, the senate, or the council of deans were completely Indonesian. It was largely outside of the official meetings, in conversations with their colleagues, that they kept up with developments and contributed their views as staff members. The progressive withdrawal of Americans from participation in meetings was a consequence not just of exclusion but of growth and development in IPB and was a testimony to Kenteam's success, even in the last months when Kenteam's feeling of being "left out" was really quite intense.

Kenteam members varied in their peripheral or central involvement in the work of their departments and in the growth of their colleagues. Keeping informed would have been much easier had they been fluent in the Indonesian language. But another relevant factor was the extent of acquaintance. When presented with rosters of names and asked how many colleagues they knew well or fairly well, some Kenteam members checked fewer than ten but one checked sixty of the Indonesian names. Acquaintance in the latter case was six times wider than average. Interacquaintance of Kenteam members was spread out in the same way; one member knew well, or fairly well, only six other Kenteam members, but another checked forty-one of the forty-seven names. Those who were most widely acquainted no doubt were best exposed to current information on developments in IPB. Spread of acquaintance was influenced by a team member's assignment: party chiefs had potentially wider contacts reaching into all the colleges; Kenteam members who stayed in Bogor for more than one tour could also have wider acquaintance. There was never a complete separation of horizontal from vertical lines of communication so far as informal interaction was involved. In official relationships, however, each Kenteam member communicated mainly with IPB associates in the department or college to which he belonged rather than directly to higher Indonesian levels.

Asked for a description of their personal relationship with IPB colleagues whom they had named as their main associates, Kenteam members mentioned ninety cases, and sixty-six of them (73%) were described as good, excellent, warm, personal, or friendly, or by some term of similar meaning. Another fourteen (16%) were mentioned as fair, reserved, cooperative, agreeable, but with no indication of personal intimacy. There were only two references to alleged non-cooperation, conflict, aloofness, and strained or distant relationships, and three in which the relationships were referred to as deteriorating over a period of time.

How this pattern of Kenteam's acquaintance would compare with the experience of other Americans in other places is not known. But their interaction with those colleagues they knew best was the chief source of current information at IPB. Often they could not understand conversations and discussions in meetings when Indonesian was spoken. Their expression of need for more complete information was only occasionally accompanied by any implication that Indonesians were deliberately withholding anything; the interpretation was rather that Indonesians understood each other and merely overlooked the fact that American colleagues were not fully briefed. Their frequent inability to find the source of authority, or information, or responsibility in relation to a given issue was a frustration for the Americans. Would the ban on sending participants to Western countries imposed in September 1964 be lifted or not? Rumor would say a few will go, none will go, all will go, they will go soon, they will go after two months. A rumor would not be confirmed or denied and there seemed to be no way but to "wait and see." The better equipped the American for full communication, the better informed he was in the course of normal association with colleagues.

Agreement and Differences

In thinking about the affiliation there is a tendency to concentrate on the differences of opinion, misunderstanding, and nonagreement, which makes them loom larger than they really should in an evaluation. Actually, during more than eight years of the affiliation, the body of agreement was substantially greater than the body of nonagreement. The selection of each member of the Kentucky team and the assignment of each Kenteam member to class and other duties was by agreement; the designation of each of the 200 participants for study in the U.S. was by agreement; the purchases of equipment were by agreement. For most of the issues on which differences of opinion were expressed at one time or another, there was eventually working agreement, an accommodation of views which allowed the work to go on while further discussion, reinforced by practice, pointed out the way toward even fuller agreement.

If and when discussion reached the stage of conflict and rivalry between factions and a taking of sides, it was Kenteam policy to remain out of the conflict. When sides were taken on who should be dean of animal husbandry, for example, or who should serve on the Dewan Curator (President Sukarno's wife, Madame Hartini, became the honorary chairman in 1965 and the Mayor of Bogor was chairman), or whether IPB should remain in or outside of the University of Indonesia, the chiefs of party insisted that these were matters entirely for Indonesian decision and not for Kenteam's "meddling." There may have been a few occasions when Kenteam members intervened in the Indonesian appointment process and promoted a department head or dean, but never without "getting burned." It was fairly well understood that decisions on Indonesian staffing were out-of-bounds for Americans.

An early and unpleasant problem was in the triangle of cliques formed by the veterinary faculty, the agricultural faculty, and Kenteam. The staff of the veterinary faculty had not been a party to the request for technical assistance through an AID contract with an American university. They had had no prior interaction with the Americans; negotiations had involved only the dean of agriculture and others higher in the University of Indonesia hierarchy and in the Ministry of Agriculture. The veterinary faculty's reception was a little testy and Kenteam response was correspondingly reserved. A complex of factors seemed to be involved: there was a tendency, which has sometimes been noted even in the U.S., for noncooperation between animal science and plant science specialists; within the animal science group there was an overprofessionalization of veterinarians which resisted the emergence and differentiation of animal husbandry as an applied field. The Dutch-trained veterinarians on the faculty insisted that only Kenteam members who were accredited veterinarians could be of help to them except with such fields as physics and chemistry in which work was required of the students in both colleges. Note was taken of the fact that the University of Kentucky had no veterinary college, but only a preveterinary curriculum in the college of agriculture. And there seemed to be a complete lack of confidence in American scientific and professional training. In addition, the intensive and productive presence of a University of California medical team at work in Djakarta with the medical faculty of the University of Indonesia provided a model to excite the veterinarians' envy.

Various efforts to heal the breach were proposed, but letters in the project files document disputes over the staffing of Kenteam, curriculum organization and teaching, the purchase of equipment, and the selection and programming of staff members for overseas training. At one climax the veterinary faculty asked to withdraw from the affiliation, seeking another especially for itself. The dispute at times required the direct attention of the director of AID in Djakarta and the presidents of the Universities of Indonesia and Kentucky. A harassed Kenteam chief of party wrote that he wished to advance his plans for going home, in part because of the unpleasantness of the crisis involving the veterinary faculty. Kenteam was not as completely or quickly successful as a change-agent with its veterinary client as with the agricultural college before 1960, except as the work in basic sciences had its effect on both colleges. In one effort at mollification, a Kenteam member was made associate chief of party to work especially with the veterinary college, but this was finally decided to be a kind of second-class-citizenship treatment of the veterinarians and was abandoned. Eventually there were changes in leadership of both Kenteam and the faculty; tension subsided and rapport was established.

The context of relationships was completely restructured, however, under pressure from higher Indonesian authorities to get veterinarians more immersed in developmental activities

42

and nation-building, and similar pressures appeared within the veterinary faculty, led by a staff member or two with supplementary American training. By 1962 the expansion of the student body, differentiation of interest within the staff itself, and the return of the first participants from the U.S. combined with other factors to establish an animal husbandry curriculum and then a marine fisheries curriculum. Thus the veterinary faculty had begun to participate intentionally in modernization. No doubt had Kenteam achieved earlier some of the skill acquired later in the empathic understanding of Indonesian colleagues, the tensions between the veterinary faculty and the Americans could have been to some extent prevented, to some extent reduced, and more or less completely dissipated.

In the interaction of Kenteam and ipb, there were many issues of discussion, some resulting in agreement, some in a diversification of views, and some in conflict followed by victory or defeat. Among the educational debates were those concerning "guided" versus "free" study, the content of curricula, the four-year American baccalaureate versus the five or more year Indonesian Insinjur or Doctor of Veterinary Medicine degrees, three-year sequences for doctorandus or sardjana degrees, general versus specialized curriculum structure, departmental organization and the principle of departmental unity: only one department in any given field, but with responsibility for teaching all students requiring instruction in that field, wherever their major location within the whole institution. Other questions were over the introduction of laboratory work, practice periods, field work, balance between basic and applied science, division of the year's time into semesters or quarters, selective admission procedures, examination practices, and student evaluation.

One area of disagreement involved differences in basic cultural values. Kenteam members, responding from their own American values, sometimes attributed problems of development to the economic attitudes of the Indonesians. There were the well-known American judgments of people in developing countries: that they assigned low status to physical work which, in the case of professors, fell to assistants or employees; that work as a technician was unrewarded by either status or pay; that traditions of gentility prevented getting "down to business"; that entrepreneurial interest and skill were very scarce; and that Indonesians neither chose to become foremen or managers nor succeeded in such roles. Other factors which some Americans met with dismay were the absence of any concepts of savings, of investment, or of deferred-gratification consumption, and ignorance of credit procedures. Americans doubted too whether the Indonesian version of mutual aid (gotong rojong), which for so long had organized manpower in villages, would be adequate to organize the complex affairs of state.

This last principle of interaction seemed to provide Indonesians with their own criteria for decisions, for example, about the use of official cars. As a "limited good," a car was judged to be needed by many people and thus could properly be used by them privately. Americans ruled that official cars could be used only officially. Among Indonesian bureaucrats, however, it was fair for anybody in a department, or for friends, to make personal and family use of a car if it was not needed at that time by the department and if the department head had granted permission, with a promise that private costs would be privately paid. At first Indonesians could not understand why Americans had to be so strict to be correct, and Americans could not see anything honest in private use of official vehicles. Had there been as many vehicles per capita in Indonesia as in the U.S., the American policy might have been understood as readily in Indonesia as at home.[1]

Another application of the principle that a good must be shared was seen in the only accepted system of rice harvest in the villages. Anyone who needed a share of the rice had an acknowledged right to enter the fields of any owner at harvest time and cut heads, one by one. The owner of the padi took his share; the harvester took hers, and her "take home pay" was large or small according to the number of

[1] Ward H. Goodenough, *Cooperation in Change* (New York: Russell Sage Foundation, 1963), p. 491.

44

harvesters. Wertheim and others have called this "shared poverty"; it was also a manifestation of Geertz's "agricultural involution."[2]

Discerning members of Kenteam noted that these cultural features, to which they gave conventional American interpretations, were embedded in Indonesian cultures and were not incongruous with the cultures of the past, yet were changing in the urgency of national development. It was out of style for Indonesian descendants from the old princely families to use the symbols of rank formerly placed before or after their names, and it was getting to be in style for both men and women to work. There were new strains toward equality, especially at loci of developmental change, such as IPB and other faculties and universities. Hierarchies of class and caste had not receded completely but neither were they immune to forces which reduced their prominence.

Another area of disagreement was the resentment from time to time in IPB that Kenteam required prospective participants to attend English language classes faithfully and to establish eligibility for study in the U.S. by satisfactory performance on language examinations. This was interpreted by some as an aspersive judgment on the candidate's study of English (one period per week during several years of the elementary and secondary years). IPB felt that whether or not a particular applicant could study in the U.S. should be an Indonesian, not an American, decision. In opposition to this view, Kenteam persistently held that proficiency in language was a *sine qua non* of successful advanced study in the U.S., and the issue was forced by the American judgment. IPB staff tried in many ways to have the language requirement modified, but never succeeded in getting Kenteam's acquiescence. Especially after the summer orientation in Lexington was established in 1962, some members of the IPB staff in Bogor felt that preliminary preparation during the academic year in Bogor, followed by travel to the U.S. and a summer in orientation, was putting too much emphasis on nonessentials,

[2] Clifford Geertz, *Agricultural Involution: The Process of Ecological Change in Indonesia* (Berkeley: University of California Press, 1963).

and that the summer period should be devoted to professional study. Emphasis on English and other orientation was never relaxed, however, and participant records in graduate schools may be submitted as evidence that the orientation was important.

Some at IPB held that too many of its staff were being trained at the University of Kentucky and that IPB would develop too narrowly because of overinfluence from one American university. The opposite criticism was made by AID, which held that since Kentucky entered into a contract for the work at Bogor, it was responsible for accepting and training all the participants. After all was said and done about 70 percent of the participants studied at the University of Kentucky, but in the total training program thirty-five institutions in twenty-nine states were involved. To reduce IPB's concern that it would be too "inbred," it was pointed out that, unlike the developing faculty at Bogor, the faculty in Lexington or at any comparable American institution had not trained its own professors, but that they were products of universities throughout the U.S. and represented a diverse rather than a homogeneous view. IPB's concern over this concentration attracted it to the idea of becoming affiliated with a consortium of universities rather than just one, if there should be another program of American technical cooperation in higher agricultural education in Indonesia. It seemed to be overlooked that spread among a large number of universities might be replaced by concentration in a smaller number, mainly the members of the consortium.

IPB staff were generally as eager to avoid involvement in Kenteam embroglios as were the Americans in keeping out of IPB factional quarrels. In several instances IPB raised questions about the qualifications of persons being considered for nomination to Kenteam membership, and in a few of these cases the nominations were not made. In the earlier years IPB was eager to have older, established men with doctoral degrees and known prestige who would be widely accepted and appreciated and who could improve the image of American scholarship in a society which respects age. Later, the insistence on age was relaxed, but IPB was always reluctant to approve men without Ph.D. credentials for membership in Kenteam (the veterinary faculty pressed for a veterinary degree).

The question of whether to request a member of Kenteam to remain in Bogor longer than his two-year assignment sometimes involved ipb and Kenteam in what was initially ambiguous discussion. The usual pattern of decision was that ipb would indicate whether the academic field should continue to be served by a Kenteam appointment. If continuation was desired, Kenteam could conclude that ipb would approve another tour for the incumbent. If a shift to another field was proposed, Kenteam could conclude that termination was being confirmed for the Kenteam member who had been serving and that a specialist in the new field should be appointed. There were cases in which reappointment was desired by a Kenteam member but without any expression of support from ipb, and cases of ipb request for continuation which Kenteam members either could not or did not wish to accept. In some cases ipb seemed to expect Kenteam to read from cues; silence or a request for change of field would normally be read as ipb disfavor with respect to extending a particular tour of duty. However, if Kenteam was uncertain of the cue and asked specifically whether ipb favored reappointment, ipb might avoid direct commitment by asserting, "It's a problem for you Americans," or by some counter question or indirect allusion, thus tossing out still another cue to be read. By terms of contract, Kenteam could not shorten a tour by sending a team member home unless there was an Indonesian indication that he was "persona non grata," yet it would be difficult—probably impossible—to get this kind of Indonesian intervention. How, by an interpretation of Indonesian-style communication, could Kenteam decide whether they were getting green or red flashes from ipb—on assignment of Kenteam members, or ordering equipment, or on selecting participants? The answer was that exceptions proved rules: some ipb members took occasional pains to speak directly and not be misunderstood. Some Kenteam members acquired skill in cue-reading. Somehow, agreement or disagreement was usually made known, decisions were reached, and action was taken, although each side leaned over backward to avoid intervention in the affairs of the other.

Staff members on foreign study assignment were sometimes recalled to resume work at IPB. Until 1965 each case was individual and not all were for the same reason. In the last year of the contract many participants were recalled. In the earlier cases, departments or deans decided that the participant concerned had "had his turn," so to speak, and could not be allowed more time than others had taken. Or situations in departments developed that required the early return of some absentees. In the 1965 wave of recall, political pressures were doubtless involved. The supervisors of graduate study in American universities were irritated in several cases, even angered, to lose promising graduate students in whom they had invested their knowledge and skill, to whom they had entrusted data, and who were in midstream toward the completion of degree work. It was difficult for IPB staff to see the reasons for this response; they viewed participant training in terms of IPB needs and could not understand the relevance of an American graduate department's concern. There were several cases in which all the wiles of an assistant campus coordinator were insufficient to keep angry graduate study directors from writing or phoning that they didn't want any more Indonesian students unless it could be guaranteed that they would be allowed to finish their work. Important graduate departments at the Universities of Minnesota, North Carolina, and Northwestern were among those alienated in 1963 by recall to Bogor of prospective Ph.D. recipients. In these decisions about recall or extension of study tour periods, it was Kenteam's policy to keep "hands off" except as each member took part within his own department in the discussion to formulate a departmental recommendation.

It was difficult for IPB staff to understand—until they or their proteges experienced it—that American graduate school regulations establish limits and that the professors who direct graduate study advise students in outlining and approving programs of study. To specify from Bogor the courses a participant should take as a candidate for an advanced degree at an American university was not possible, but this was seldom understood at IPB. The faculty there and the participants usually had conferred in Bogor about the kinds of training needed, but graduate department endorsement in the United States of these plans was difficult to arrange. Personnel in the office of the campus coordinator in Lexington invested much time in trying to reconcile the recommendations from Bogor departments and Kenteam professors with the requirements of graduate departments and study supervisors.

The strategies employed by Kenteam in dealing with issues that IPB faced in its organization and educational development were posited on the recognition that IPB was an Indonesian institution with its own goals and its own destiny, to be fulfilled in its own social and economic context. The aim of Kenteam was to lend the expertise of its members and the resources at its command toward the implementation of IPB objectives, and in no sense to impose other concepts and systems. Whether or not IPB would build strength in food technology; whether it would embrace and integrate the fields of marine and inland fisheries; whether it would emphasize animal science in other than veterinary terms; whether it would offer specialization in biological science, agricultural processing, home economics, or forestry; whether it would seek strength in research capacity; whether it would engage directly in extension and public service or leave work in these fields to the responsibility of other institutions—these and many similar matters were questions for Indonesian decision. Kenteam members, each in the context of his relationships with departmental colleagues, were resource persons and consultants, helping to identify alternatives and to foresee the consequences of action in various possible directions. IPB studied the issues often in its own meetings and its own language; IPB members and Kenteam members discussed them in various conversations, two by two, in departmental meetings, and in committees. Kenteam members discussed the issues in team meetings and informally in numerous conversations. But Kenteam did not vote, pass resolutions, or assert any team positions.

The most severe criticism of Kenteam ever made by Indonesians came in a memorandum responding to Kenteam's SAD memo—a statement on "Study for Advanced Degrees," which set off a wave of resentment against Kenteam and triggered resolutions from a student organization to President Sukarno demanding that Kenteam be banished from Indonesia. The statement, widely thought by Indonesians to deprecate the Ir. degree, precipitated Kenteam's greatest crisis, and Kenteam survived only by a formal expression of regret from the chief of party. The IPB criticism argued that Kenteam, having worked in Bogor for six years, was at least jointly responsible for remaining deficiencies at IPB, that Kenteam members could have worked in departments to spark more constructive thinking on curriculum; that there was insufficient interaction within the affiliation; and that Kenteam members on graduate committees had not made enough contribution and should have influenced better scholarship on the part of Indonesians in Bogor.

Kenteam Restraint

Kenteam kept itself voluntarily more or less housebound, or at least restricted to campus and neighborhood, at certain times of crisis: the Lumumba affair, the Cuban crisis, anti-Chinese riots, and Ash Wednesday in 1963, when the British embassy was burned. This restriction was as much to prevent aggravation of unrest as for personal security, for which Kenteam members seldom if ever feared. There were a few occasions when their houses were under guard by special arrangement of IPB executives with the army. The Indonesian-American ladies' group could not call itself a club without being approved, and this would have been too much to expect. American children were kept up to date by a community tutoring program but could not call it a school without violating an Indonesian proscription of foreign schools.

Kenteam's restraints, mainly self-imposed and only rarely requested by IPB colleagues, followed changing patterns and were lifted and reimposed as circumstances seemed to require. In the early years Kenteam members wore their academic caps and gowns at appropriate ceremonies to reenforce their own credibility and IPB's image, to lend color and ritual to IPB's growing sense of identity, and to dramatize higher education. By 1963, however, it was necessary to burden the IPB-Kenteam grapevine with a request that the colorful concentration of Americans be unrobed and scattered in the general audiences, rather than seated with the faculty, to reduce visibility at a commencement convocation. In the early years, Indonesians took evident pride in the presence of American professors behind the lectern, in the departmental offices and meetings, on the stage, or among the VIPs at cultural and educational events. Later, not because of changes at IPB but in response to the changing political environment, the absence of Americans was preferred to their presence. In ceremonial examinations of candidates for the Ph.D. degree, however, American professors continued to serve as co-promoters, to participate in the questioning, and to wear the robes of academic sanction.

Courtesy imposed restraint on the Americans in discussing their relative affluence and their successes in the souvenir markets with less privileged neighbors. The cost of purchased items was always of great interest to Kenteam's families, but they lived still mainly in a dollar economy and participated only partially in the rupiah economy of Indonesia. Abstention from gossiping with Indonesians and from voicing derogatory political and economic comments was also self-imposed in courtesy as well as wisdom. There were, of course, many slips of the tongue and occasional breaches of good taste, but consciousness of the restraints generally prevailed.

Sometimes a slip of a tongue revealed that some aspect of a relationship was slightly out of adjustment. When a Kenteam member referred to the study of educational needs at other faculties as an AID survey he was not only wrong in terms of fact but insensitive in terms of relationship. The survey in question was made by the Ministry of Education, managed by an Indonesian committee, and should have been so explained, although Kenteam members were on the survey teams and AID financed the travel. The Indonesians who travelled with the survey teams explained to some of their hosts at the colleges visited that it was the ministry's survey and that the Americans had come along to take advantage of the opportunity to travel. From Kenteam's viewpoint it was important that both the relationship and the language of reference to it should properly acknowledge both responsibility and recognition.

Only toward the end of Kenteam's stay did the pressures for restraint steadily tighten. In the final weeks, Kenteam members kept their offices in their homes rather than in the departments. They did not meet classes, were consultants instead of teachers, and to be sure of a meeting with the rector, the chief of party went to his house before 7:00 A.M.

Concluding Note

Several items in this review indicate Kenteam's problems of communicating within the proprieties of Indonesian discourse, and conversely, IPB's problem of accommodating to the directness and immediacy of American expression, whether written or spoken. There was an unstable balance between misunderstanding and understanding, but clearly enough of the latter to keep things going and a continuous reciprocal effort to enlarge it. In general, Kenteam members tried to understand Indonesian styles in communication and learn and practice the American manner of direct expression. The result was the kind of accommodation that mediates communication between peoples who are learning each other's language more by working together than by formal study. No doubt IPB's gesture toward accommodation was greater than Kenteam's because most of the Indonesians were seriously learning English and most of the Americans were not seriously learning Indonesian. There was equality in the pressures involved, of course, because in learning English the Indonesians gained permanent access to world science and technology, and most of the Americans could anticipate only a few months or at most a few years during which they to modify their own signals, while Indonesians sought to

A second major grouping of problems—and it could really would—if they could—use Indonesian.

be subsumed under communication or set out separately for visibility—is that of values, even though this concept is vague and imprecise. Several of the foregoing comments on the relationship between Kenteam and IPB concern their problems in understanding each other's traditional values. A reason for treating the two problems together is that when there was full communication, value differences were minimized and tolerance of differences was maximized.

Most of Kenteam's problems in relating to IPB involved ambiguities of role. It was not crystal clear which model should prevail: Were Kenteam members and IPB staff to face each other as guest and host, agent and client, advisor and advisee, donor and recipient, leader and follower, senior and junior, expert and layman, or colleagues, counterparts? All of these models were used, each Kenteam member finding a manner of association which he chose, or believed appropriate and feasible, or had thrust upon him by circumstances.

Of all these models, however, the one that emerged from Kenteam's experience as best suited to the joint effort in institution-building at Bogor was the host-guest relationship.

Perhaps Kenteam's greatest problem was in achieving the view that their Indonesian colleagues were hosts, not clients. And this was the converse of IPB's problem: receiving the Americans as guest professors and not agents of intervention.

There were strains in the guest-host model and many lapses from it, but Kenteam members were generally acknowledged to be guest professors and generally fulfilled this prototype role.

3. Ground Rules

Some of the rules by which the University of Kentucky team lived in Bogor were unique to the time, the place, and the persons involved. But the problems they solved would appear in any similar undertaking.

The formal framework was the contract which provided for a three-part program: technical assistance (the work of Kenteam members); procurement of educational materials (commodities); and advanced study in the United States by staff members of IPB (the participant program).[1] In addition to procedures generally required by the contract, a body of practice accumulated in response to special needs as they appeared, and these were the ground rules.

The contract and the ground rules were to each other as statutory law is to common law, and every similar institution-building project will need the same duality of regulation. The ground rules guiding Kenteam were in part applicable only to its American members, but they were also a sub-set within a larger structure of regulations that governed the whole of IPB. Emphasis in this evaluation of Kenteam is on rules that applied mainly to the self-regulation of the Americans and were mainly generated by them.

Procedures for Choosing Kenteam's Members

The first step in determining the team's composition was the identification by IPB leadership of fields in which help was needed. Thereafter, the deans (in later years, the rector) and Kenteam's chief of party sent job descriptions to the campus coordinator, who sought candidates. The initial staffing pattern was written into the contract and modifications in later years were authorized by contract amendments. First priority was given to any qualified and available regular staff member at the University of Kentucky. Only 40 percent of all Kenteam members, however, came from the university staff.

54

The qualifications of candidates were then studied by both IPB and Kenteam and questions were often asked. Adjustments were usually necessary; in hardly any case did the available nominee fit perfectly the described need. The appointment of one animal husbandryman, for example, was delayed because he was not a specialist in genetics. Some were questioned because of age. (Initially IPB thought older men would command more respect; later this preference was abandoned and competence was stressed over age.) A few nominations were questioned because the nominees lacked the Ph.D., which IPB considered highly desirable. It was questioned that one biologist might be too narrowly specialized. It was questioned that one nominee's experience was in extension rather than teaching. Some nominees were withdrawn because of doubts expressed by IPB or Kenteam members. But all who served in Bogor survived the gamut of approval by IPB, Kenteam, the University of Kentucky, and AID in Washington and in Djakarta. The normal period of service was two years but some terms were longer. There were Kenteam members who sought reappointment; among these, some were and some were not approved by IPB for extension.

The first members to serve in Bogor had no part in selecting each other but they did influence the selection of their successors. The last to serve had no successors to select. Hence, when questioned later, half of the Kenteam members (51%) said they felt they had as much influence as they wanted to have on the composition of the team. There were a few, however, (30%) who felt they had *not* had enough influence, stating that they generally had had no opportunity to express an opinion, or were not consulted, or that AID intervened, or that IPB "didn't plan with us skillfully enough," or that the chief of party "seemed to have charge of this." It finally came to be established in the middle and later years that needs were

---

[1] A later standard form for AID-university contracts was better suited to fulfill the broad educational interests and obligations of universities. See John W. Gardner, *AID and the Universities: A Report from Education and World Affairs in Cooperation with the Agency for International Development* (New York, 1964).

indicated by IPB during and following joint planning with Kenteam, followed by full discussion within Kenteam.

The appointment and work of each Kenteam member was to some extent independent of that of every other member. There were some, however, who preceded or followed others in the same field or a closely related one. For example, a botanist was succeeded by a plant physiologist, a forester by a second forester, the first virologist by another. Excluding administrative officers and technical secretaries, 65 percent of the members of Kenteam were either predecessors or successors of others. In some areas, arrangements for additional work were made by reappointing a Kenteam member for another term. In cases of succession, however, and excluding administrative officers and chiefs of party, there was no instance of overlap, and predecessors left Bogor before their successors arrived. This resulted in discontinuity and loss of momentum and prevented effective follow-through on some work which had been undertaken. Nearly every team member finished his term as a lame duck. The cumulative impact of Kenteam at Bogor could have been greater if there had been smoother administration of the succession of outgoing and incoming members.

Kenteam as a Team

The unit of action in carrying out the purposes of Kenteam was the individual member, not the team as a group. Nevertheless, the Americans required some organization of themselves as a team. They needed to help each other with orientation and understanding, to discuss their work, review problems, identify strategies for better accomplishment, and administer the logistics of living and working in Indonesia. Hence, Kenteam was a discussion group, an interest group, and an association of colleagues in work.

As time passed, Kenteam became somewhat integrated as a special functional unit of IPB. Not more than two or three times in nine years did it take parliamentary action by voting on motions or acting by resolution and then not on academic concerns of IPB. It maintained active special committees from the beginning, however, and its consensus or division of opinion on many issues was achieved through discussion. Team views and judgments were formulated and they were interpreted by the chief of party to IPB deans and rectors, to AID/Djakarta, and to the University of Kentucky. There was a great deal of interaction between Kenteam members in their family and neighborhood activities; their problems and plans became well known without mechanisms of formal communication.

Kenteam resisted organization, yet inevitably became organized. In a sense, it was an American clique within an Indonesian university. Sub-cliques also appeared informally from time to time within the American group. There were periods when members who belonged regularly to the campus staff of the University of Kentucky were thought by others to be a clique. There were periods in which those who belonged to the veterinary faculty were regarded as a clique, and when those who had young children were thought by others to be a clique. These groupings were strictly informal but sometimes their interests and influence resulted in new ground rules or modifications of old rules.

Kenteam was a team, it had a chief, an administrative officer, a staff with personnel serving as clerks, secretaries, accountants, drivers, messengers, interpreters and translators, "trouble shooters." It held meetings, formed committees, prepared reports. Hence, while Kenteam strove to lose itself in IPB, it could not avoid becoming a body both within and apart from IPB.

The Administrative Staff

To mediate the contract and the ground rules for Kenteam, the special roles of chief of party and administrative officer were established, and Kenteam had successions of three persons in each of these positions.

The position of chief of party has no exact analogue in the organizational structure of an American university. It is similar in some respects to the chairmanship of a large department, but specialization among team members ranges widely among many departmental fields of interest. Yet it is not like the position of the dean who presides over a complex of fully staffed departments.

The unique features of team leadership include responsibility for a balance of authority and accountability, initiative and response at a special nexus of communication through which messages flow from and to the home university, the host institution, colleagues on the team, agencies of the American government, agencies of the host government, and numerous elements in the host community. None of the chiefs of party in Bogor ever saw a job description that would resemble the duties they performed, and each identified the character of his work by mixing experience and judgment in a new crucible of practice. Leaders of overseas teams are an emerging profession of educator-administrators, special in function but probably never to be numerous. Although vested with authority by the contract and all parties to it, the chiefs of party during most of the Kenteam years in Bogor conceived of their role as chairman-of-the-group rather than as an authoritarian leader.

Kenteam's view of the personal and other characteristics essential for a chief of party's success is expressed in a composite list compiled by content analysis from statements by team members. The following is a nontechnical formulation and is couched in the evaluative adjectives and phrases of common speech; the characteristics do not classify easily and they compose a catalog of virtues, here grouped under five headings:

The first requirement is for *"understanding of the task"* and *commitment to it.* Kenteam stressed realization that the transcendent purpose of technical cooperation is to help a host institution, and a team leader can serve this purpose only if he has "a broad outlook"—an ability to see the total picture and properly evaluate each component. Included also is the conventional American requirement of any leader, "willingness to work above and beyond regular schedule."

Next in order is a requirement for *competence* in several dimensions: "broad technical competence as an academic man." (Kenteam's first chief of party was a geneticist, the

58

second a chemist, the third a rural sociologist.) Need for knowledge of the host country's language was stressed. The concept of competence embraced knowledge and understanding of—and ability to exemplify—American culture and the United States as a nation. It embraced also interest in and understanding of the host society's ways of life, and ability to relate effectively to the host society and its officials.

Administrative capacity was specified and spelled out in various ways. A team leader should be able to plan, to decide, to delegate authority. He should be firm but also flexible in exercise of leadership, should have self-confidence (but not overconfidence), patience, perseverence, ability to communicate well, and ability "to spot phonies!"

Of equal importance with the foregoing attributes, and overlapping them in some ways, is the requirement that a chief of party be *supportive of colleagues*. Kenteam felt that a team leader should desire the success of colleagues, recognize accomplishments, and extend praise therefor; that he should be possessed of natural empathy; that he must be able to identify and respect diversity of viewpoint. There is a quality of "father-confessor" in the Kenteam concept of the role. Phrases used in the description of this broad requirement include: respect for others ("treating adults as adults"), desire to help, impartiality, sympathy, cooperativeness, enthusiasm. There was stress on working with colleagues to make each field important, ability to get along well with others, understanding that "overfed Americans" have problems, too, having a "willing ear" for any team member, and being sensitively considerate of the families of team members.

A final set of requisites refer to *character* in general, supplementing the priorities indicated above by such further specifications as sincerity, tact, loyalty, sense of humor, high ethical standards, rejection of gossip and graft, objectivity, and freedom from bias.

Three administrative officers also held office successively during Kenteam's stay in Bogor. The division of labor between the chief of party and the administrative officer was worked out in the field rather than fully prespecified in job descriptions. In broad distinction, the chief of party worked at policy formulation, programming, and mediation of relationships. The administrative officer was in charge of the logistical implementation of the Kenteam task. His chief duties included general organization and supervision of office staff, processing arrangements for IPB staff members chosen to study abroad as participants, processing equipment orders, clearing goods through customs, maintaining housing and utility services, receiving and installing newly arrived families, assisting terminating families to prepare for departure, and general "trouble-shooting." Had these tasks been the direct responsibility of the chief of party, they would have distracted his attention from planning and mediating. They were tasks which indigenous personnel at IPB were not yet fully prepared to perform and tasks which other American personnel in Djakarta would have found difficult to execute in behalf of Kenteam and IPB.

Kenteam members recorded a long list of personal and other characteristics they thought essential for an administrative officer's success. To some extent these overlap and are identical with the bill of particulars for the position of chief of party, yet they seem to group naturally in a slightly different manner.

As in the list of requisites for the chief of party, understanding is mentioned first—though with a slightly different emphasis: "*understanding and acceptance of roles.*" This means understanding and accepting the mission of the whole enterprise and the difference between his own responsibilities and those of the chief of party. An administrative officer's understanding and acceptance of his role was thought by Kenteam to involve a genuine desire to be helpful to all parties to the project; a willingness to keep out of the chief of party's policy role; an interest in routine performance as well as in meeting emergencies.

By *administrative competence,* Kenteam meant that an administrative officer should be a "good organizer"; should know how to set up and operate office functions; should know intimately the terms of the contract; should be fully communicative and tactful, yet decisive; and should have a strong sense of responsibility and loyalty.

Skill in relationships with others is a third broad category of requisites identified by Kenteam members. An administrative officer must "work easily" with different people under difficult conditions; he must be respectful, friendly, a "peacemaker" who is fair, firm, considerate, diplomatic; he must be able to "keep an ear to the ground" and advise the chief of party; he must employ "leadership abilities" and have recourse to an appropriate sense of humor.

Qualities of personality are indicated in all the foregoing paragraphs, but Kenteam had other traits in mind, as well: patience, poise, tolerance, honesty, energy, cooperativeness, courage, dependability, self-control, political know-how.

Until the last year, a room in the university's administrative building housed an office, identified by a sign above the door reading "Kentucky Contract Team." It was moved across the street into a vacated Kenteam house in the last year under the double influence of IPB expansion and waning solidarity in the relationship with IPB as a result of "the situation" and approaching withdrawal. Here was the lair of the administrative officer and Kenteam's administrative staff, which included one American woman serving as technical secretary, and at first two and finally five Indonesian women with clerical, stenographic, accounting, and secretarial roles. Kenteam's Indonesian secretarial personnel became one of the strongest instruments of coordination throughout the Kenteam period, persisted strong and unbroken through all the periods of tribulation, and never ruptured even at the moment of Kenteam's withdrawal. The Bogor staff employed also one Indonesian man as a messenger, three to five as drivers, and five to eight as night watchmen for Kenteam houses.

Could these persons have been employees of IPB, performing the same functions under IPB control as they did from a Kenteam administrative office? Should they have been somehow integrated into IPB administration as Kenteam members were integrated professionally into the academic departments? The terms of the contract and its international dimensions seemed to require some parallel—but separate rather than integrated—administrative organization. American laws, agencies, personnel, and customs were involved as well as, and to some extent apart from, Indonesian laws, agencies, personnel, and customs. It is doubtful then whether IPB could have organized the facilities which the Kentucky office was able to provide.

One of the problems in a cross-cultural project like Kenteam's is the inevitable marginality of local staff. The community may come to regard its members with a special aloofness, mingling jealousy for their favored status with suspicion that they exploit the opportunity to be close to Americans. They may be thought by some to curry favor and monopolize friendships with Americans, especially by helping them solve problems and by mediating their comforts. There is the possibility that some members of the community may see superiority and arrogance in their manner. Kenteam staff got higher pay than many IPB professional staff; they rode around in team cars on team business, became close friends and favorites of Kenteam wives, served as confidants, advisers, interpreters, trouble shooters, representatives, and so on through a series of cross-cultural roles. It was they, and to some extent the counterparts of the American professors, who nurtured a "third culture"—a kind of sweet-and-sour culture (to borrow a homely figure from the Chinese menu) in which the ways of Indonesians and Americans were sorted out and accommodated and integrated in pragmatic blends.

Communication

Communication between the parties involved in the Kenteam project occurred within a network that one chief of party figuratively called a game of puss-in-the-corner, in which five parties competed for space in four corners: AID Washington, the University of Kentucky, the government of Indonesia, AID Djakarta, and Bogor (IPB).[2] Actually this oversimplifies the situation; each of the five parties was in itself more than one, as the Indonesian government with its

2 See CIC-AID Rural Development Research Project, *Building Institutions to Serve Agriculture* (Lafayette, Indiana: Purdue University, Committee on Institutional Cooperation, 1968), pp. 213-15.

ministries, or Bogor, with the bilateral American and Indonesian presence at IPB. Certain channels of intercommunication were designated in the contract and others evolved as ground rules at the site of work.

Appropriate channels were not all identified in the early period nor yet in the middle period, and there were occasional moments of tension, disappointment, even anger, as a result of uncertainties and violations flowing more from inexperience than from intent to maneuver. With experience, however, Kenteam identified its proper connections in the network of communication. One problem was that, taken all together, the parties to the contract-affiliation were members of a complex organization, and communication within the complex organization had to be regularized by bureaucratic procedure, not always welcomed by persons at work within the bureaucratic system.

The basic tools of bureaucratic communication—typewriters, copy machines, mimeograph, calculator, and files—were at hand. The old rule that each communication should deal with only one topic was asserted and more or less universally and chronically violated, but appeased by devices of filing and referral: a copy of multi-topic letters and memoranda to each interested party and a copy in each relevant file. During the first five years, the chiefs of party kept their own private files, apart from the general team files, but during the last four years there were only the general files. A special confidential file was kept by the administrative officer in behalf of the team. Trunk lines of formal communication were eventually recognized: on the American side they were between the campus coordinator in Lexington and the chief of party in Bogor; between the campus coordinator in Lexington and the AID contract officer in Washington; and between the chief of party in Bogor and the AID contract representative in Djakarta. Staff members in each location used only these "main trunks."

In the functioning of this arrangement, however, certain strains developed. One involved the physical means of intercommunication. International telephone was used occasionally but never conveniently in the early years, and finally not at all. However, the number of telegrams exchanged increased considerably and the volume of correspondence flowing both ways seemed to increase. International mail was never completely dependable and internal mail was exceedingly slow. Correspondence was greatly facilitated by extension of APO eligibility to Kenteam during the last years (1962–1966). Within Bogor and between Bogor and Bandung or Bogor and Djakarta, messages were generally hand carried.

One strain was sociological in character—the pressure of personal relations on impersonal rules. Kenteam members were in personal touch with each other, with their IPB colleagues, with some AID personnel in Djakarta, with participants studying in the U.S., with staff members in the campus coordinator's office, with university faculty in Lexington, and even to some extent with AID officials in Washington. The context of personal interaction demands freedom for intimate intercommunication between friends, so "shadow" networks exist, unseen within formal bureaucratic networks. Sometimes the formal and informal systems interfered with each other.

These may be old problems with familiar solutions to career bureaucrats or career diplomats, but many professors are neither, and a lesson from Kenteam in Bogor is that systems of communication are important enough to require some prearrangement and continual self-monitoring by those involved.

Another dimension of cross-cultural communication at Bogor was the need to take account of the traditionally embedded yet rapidly evolving status structure among Indonesian colleagues. Revolutionary norms of equality were penetrating Indonesian society. The prince and the commoner were together as colleagues in a department, and new norms of status were appearing to conform to the degrees and grades of service in the struggle for independence, in the hierarchies of professional rank, service rank, and administrative control. The criteria of rank and status were changing but the sensitivity to position remained strong. The ministry and various ministry officials, the rector, associate rectors, deans, associate deans, heads of departments, professors, faculty members with and without doctoral degrees, part-time staff, students, personnel in supporting services—all were potential recipients of communications, and Kenteam had to learn, therefore, what matters were for communication to whom and at what level.

Committees

Some committees were formed for matters completely internal to Kenteam. Two of the chiefs of party appointed and presided over policy advisory committees of three to five members. In some years there were social committees, committees to assign houses, to arrange lecture and discussion meetings for team members, to formulate ground rules for transportation, to distribute surplus foods to IPB students, to seek sources of research funds. One committee, including some Americans who were not Kenteam members, served as a Bogor-American school board. Internal administrative committees were in charge when chiefs of party were absent for home leave. Each committee tended to contribute to Kenteam's esprit de corps and thus to increase the extent to which American problems were solved by Americans. They may, however, have prevented fuller absorption of Kenteam into the Indonesian institution and community.

The question of whether functional committees should be bilateral was answered in different ways at different times and with respect to different concerns. The participant committee and the commodity-purchasing committees were really IPB rather than Kenteam units, and they came to be jointly appointed by the dean or rector and the chief of party, and with both Kenteam and IPB membership. The responsibility of the participant committee was to arrange the selection, preparation, and sending abroad of staff members to engage in advanced study, preparing for later service in preplanned positions on the IPB staff. The responsibility of the commodities committee was to screen requests from the departments for the purchase, from dollar funds under the contract, of needed educational equipment. Occasionally, but not for long, there were bilateral curriculum committees, appointed by the dean or rector and the chief of party.

Meetings

Practices with respect to Kenteam meetings varied; throughout the period there were more or less frequent and regular meetings but no recurring pattern of regularity. (At one moment of crisis near the end there were three meetings in one day!) Members would gather by arrangement in the Kenteam office or occasionally in one of the homes. Minutes or notes were generally, but not always, kept and filed. The usual meeting procedure was to "go around the circle," allowing for an expression from each member according to his interest or work at the moment. Occasionally an advance agenda list was made ready by the chief of party or administrative officer. Committee reports were heard and discussed, as were suggestions, grievances, and problems. Only once was a joint meeting held with the Kentucky team at Bandung. Occasionally but not usually guests from AID or other agencies were present. Occasionally senior personnel from IPB were present. Occasionally the minister of higher education, the rector or the dean would ask—or be asked—to attend. Occasionally a meeting or a series of meetings would be held to get team thinking crystallized and summarized on some issue or topic. Sometimes wives attended meetings.

An illustrative outcome of team discussion was a memorandum entitled "The Role of Agricultural Faculties in Developing Countries," requested by IPB officials for use in a national conference on higher education in 1963. This was distributed by AID in Washington to its numerous missions around the world, and later was published in a collection of readings on agricultural development.[3] Another illustration was the memorandum on *Study for Advanced Degrees*, which became an occasion for some controversy (described above, pp. 25-26, 49).

The subjects discussed were varied. There were disaffections rising sometimes to the level of minor quarreling in Kenteam and the ground rules were occasionally adjusted in accommodation. For some cases, ground rules were irrelevant.

[3] Raymond E. Borton, ed., *Selected Readings to Accompany "Getting Agriculture Moving,"* (New York: The Agricultural Development Council, 1966), 2:839-44.

Was it "right" for an unmarried member of the team or a couple without children in Bogor to have the same housing as a family with children? The Indonesians—and some Kenteam members—thought not, but each single person or couple stood his or their ground by U.S. standards and in every case had equal housing. There were misunderstandings over school arrangements for team children, charges of neglect by the administrative office in procurement of kerosene, tussles over the allocation of furniture for the houses, and allegations of inattention to needs for housing repair. It was sometimes even charged that a chief of party cared more about the Indonesians than about "his own people."

Conducted without rules of order, often without pre-arranged agenda and with discussion-circle rather than parliamentary procedure, Kenteam meetings enhanced solidarity, identified and resolved problems, and allowed Kenteam members to express themselves more fully than they could in any Indonesian meeting. Meetings occasionally gave members sanctuary from troublesome or frustrating moments, provided a safety-valve for expressions of irritation and for aggression directed toward AID, UK, IPB, or the Republic of Indonesia, provided them with a continuous planning interest, and encouraged them to keep on with—or modify—procedures for their work in the departments. Meetings were also sometimes a target for the American-style criticism that time was being wasted! And although it might seem that Kenteam members would have found satiety in each other's company, most members attended most meetings and there is little doubt that they usually got somewhat recharged by the discussions which went on.

Reports

Several kinds of written reports were submitted on the work of Kenteam and its members. Initially, reporting began as a contractually-required, bureaucratic, routine accounting of time and money expended. Reporting approached, but never achieved, full potential as an exercise in evaluation and planning. A quarterly report on personnel and disbursements was prepared by the office of the campus coordinator in

Lexington and submitted to AID. Reports were written annually by visiting inspectors from the home campus. Team members submitted annual reports, reports of travel, special occasional reports, and terminal reports. Initially the reports were addressed to the chief of party for transmission to the University of Kentucky headquarters. Later the idea was accepted that team member reports might properly be directed to their department heads and deans, and supplemental items of concern especially to American administration were directed separately to the chief of party. No standard format was developed; each report was to be its author's self-evaluation in the context of his own working situation and an analysis of emerging problems with suggested directions of future effort.

Kenteam reports veered toward one or the other of two basic types: recommendations and reports-of-work-done. Many KCT members felt the urge to recommend. Sometimes the tone was politely imperious and the verbs were buttressed by "ought," "should," or even sometimes "must." Others let recommendation be implied in reports of activities going on. The plan-of-work report was usually in expository style and dealt with what was being done or with what would be done.

The stance of the writer was that of an IPB department member, and the purpose was to stimulate, facilitate, and contribute to departmental, college, and university planning. Preparation of the reports was preceded by discussion with Indonesian colleagues. Distribution of the mimeographed reports was through the chief of party to the deans and rector, to AID officials in Djakarta, and to professional colleagues of the author, whether in Indonesia or elsewhere. Report writing in a cross-cultural context imposes restraints on taste in expression to avoid political involvement or offense, and there were one or two teapot-tempests following the distribution of provocative documents which occasioned some regret. Kenteam outgrew the minimum requirement of report writing for accountability but never achieved full exploitation of report writing as a teaching and counseling technique and as a means of educational planning.

Routine Administration

Whether the main business of the Kenteam office was to execute the administrative terms of the contract or to serve the professional and personal needs of Kenteam members and families was sometimes debated among the Americans in Bogor. The first view was more likely to be that of the administrative officer, the second more likely to be that of a Kenteam member in Bogor. It was a matter of disagreement, for example, whether to mimeograph teaching materials for Kenteam professors in the Kenteam office or to leave such service for the IPB departments in which the professors were at work.

A factor in the security of Americans overseas is the provision for emergency travel, and ten out of the forty-seven Kenteam families, numbering 137 persons on post, had occasion to invoke this provision. In every case but one the emergency related to the health or family responsibility of a team member's wife. Another category of emergency travel was that occasioned in 1965 during days of political insecurity and tense anti-American feeling by the ambassador's order that women and children in official American families be evacuated—an order which was in force for a period of six weeks.

Routine travel within Indonesia required the formulation of numerous ground rules. Kenteam leaders and IPB administrators recommended that official trips be made with the company of an Indonesian colleague, either the highest level faculty member available to go along or another to take his place. Kenteam members sometimes travelled with their families and usually but not always with Indonesian colleagues and friends. An examination of trip reports indicates that about one-third were made without Indonesians (except usually an Indonesian driver, not mentioned in the report but taken for granted). Two-thirds were with Indonesian colleagues (students, associates) and friends.

There were several reasons, different in nature and in order of importance, for having Indonesian companions, not just a driver or servant, when travelling internally. The best reason, perhaps, was to foster cross-cultural acquaintance and friendship; others were to have a guide/teacher/interpreter of the culture and language; to provide opportunities, otherwise not available, for colleagues to travel; to provide opportunities in which Kenteam members could give technical instruction as consultants to cotravellers. A final factor which must be acknowledged is that travel in some areas in the early years was restricted by curfew and state-of-war regulations. Convenience was always a factor, also, and help in finding lodging or ordering food and drink. Occasionally Kenteam members and their families travelled without Indonesian colleagues in disregard of these factors in order to demonstrate their independence and capability, to save costs, for privacy, or for various other personal reasons.

One source of difficulty in administration at the Bogor end arose over changes in contract limits or changes in interpretations of contract provisions. Many of these engendered disappointment and unrest among Kenteam members; they sometimes felt their chiefs of party and campus coordinators were failing to represent the interests of Kenteam in the face of bureaucratic policy change. Certain changes in facilities, on the other hand, such as the provision of embassy health and medical services, inclusion in U.S. commissary and APO privileges, and the increase in educational allowances for children greatly buoyed Kenteam morale when they were made available in the latter half of the period in Bogor.

Privileges of import, exchange, communication, and purchasing could not be used in behalf of others. These restrictions came to be accepted and followed but were always a source of embarrassment to some team members, who wanted to do things for Indonesian friends. These restrictions came to be understood also by the Indonesians but with some effort on their part because of their traditions of equality and sharing. It was difficult for them to accept, for example, the fact that Kenteam official cars were only for Kenteam members to use. *Their* official cars were for everybody—even for personal use, if permission was granted by superiors.

Specific ground rules of administration were numerous, as illustrated by the following list:

Team cars are for official use, approved by the administrative officer, and not for personal use.

A newly arrived family may use a Kenteam car and driver until the family's personal car arrives.

Travel to Djakarta for health reasons is official.

A car and driver on official travel in Djakarta may be stopped for only fifteen minutes for personal errands.

Requests for housing repair or other service from the Kenteam office must be submitted in writing.

Office staff members will help new families initially to install and instruct servants. Thereafter, they can help only by special arrangement.

Each Kenteam family will be assessed a monthly contribution to pay supplementary wages to night watchmen.

Kerosene deliveries will be by ration and delivery schedule.

Each Kenteam family will entertain every newly arrived family as soon as possible.

A "survival kit" of household equipment, belonging to Kenteam, will be assigned to each new family for use pending arrival of the family's shipment.

Solicitors calling at homes for contributions [and there were many!] may be referred to the Kenteam office for clearance with Indonesian staff.

One of the responsibilities of the administrative officer was to oversee the housing of Kenteam members. The period of maximum team population, 1962 and 1963, required more housing space than the twelve dwellings available (nine on one street, three on another) near the university. There was a spillover into houses of another style and farther from the campus. This issue aroused some quarrelsome discussion. The temptation was to conclude that Kenteam's size should never again exceed the capacity of the twelve houses identically built and arranged. Counter to this, however, was the view that program and need at IPB and resources available to AID should determine Kenteam's size and that availability of housing could be arranged accordingly. It happened that a decline in the number of Kenteam members alleviated the problem. In each generation of Kenteam members in Bogor, however, there were some who regretted being housed in adjacent and identical properties rather than in scattered and diverse homes in the community. Kenteam's experience indicates that it would have been better if Kenteam families could have been dispersed in the community to live as neighbors among their Indonesian colleagues, just as their offices were dispersed in the departments among their professional associates.

Planning and Budgeting

The annual visits from University of Kentucky inspectors were a stimulus to faculty planners. Usually a visiting official from Lexington was scheduled to meet with each dean and department head for discussion of recent activities, current needs, and expectations. The exercise of getting information assembled for those meetings was a learning experience and it no doubt also strengthened the esprit de corps in the sections, departments, and colleges, as they reached agreement on what to do and say in the planning meetings.

All parties expected that negotiation to designate areas of need for American personnel would initiate in IPB with Kenteam participation and would be reported with "job specifications" to UKOOP for planning the scope of programs and undertaking the assignment or recruitment of professors. AID funding was then expected to be in terms of the planned dimensions reported from Kenteam.

AID required Kenteam plans approximately annually, and bureaucratic deadlines were so timed as to require quick Kenteam response for budgeting purposes. AID plans were more or less in disregard of whether there had been any IPB participation in planning. It was a strain for Kenteam to work with IPB on planning and equally so to get UK and AID to recognize the necessity for local focus and initiative in planning for IPB.

Kenteam's ground rules for planning and budgeting never kept it far enough ahead of AID's interest—or UK's—in determining the scope of the program for future periods. Volumes of letters between Lexington and Bogor dealt with the budget. However, the impression grows as one reads the correspondence that there was a lack of clarity with respect to the principles of budgeting, particularly with respect to the allocation of responsibility for budgeting, and most particularly with the lack of preliminary planning of programs for which budgets should provide funding.

Because it wasn't far enough ahead to take the initiative, Kenteam was sometimes in the position of responding to pressures from Djakarta and Lexington. Kenteam's unfortunate position in this matter resulted from a lag in formulating procedures, awkwardness in getting IPB to undertake the necessary review and planning activities, IPB's still unformulated aims and goals, the short periods of service by Kenteam in Bogor, general uncertainty in both IPB and Kenteam about the character of Indonesian agriculture and its proper future role in national development, and the general hesitancy of IPB to propose and Kenteam to impose. In any such project as Kenteam at Bogor it is imperative that every possible priority be given to initial and continuing planning by the developing institution, with participation of guest specialists, and for submission through the channels of approval before limits are announced downward and backward through those channels. Kenteam did not have the influence on plans and budgets that it should have exercised because it couldn't stimulate enough early initiative from IPB and was always in the position of having to react to what AID and UK said would be the limits in provision of equipment and supplies and in support of advanced study in the U.S.

The Commodity Program

Agricultural faculties everywhere need land and buildings. When Kenteam first went to Bogor, IPB had hardly more than a front and back yard. A few outlying tracts were added from government acquisitions during colonial withdrawal. Land and buildings were not a direct responsibility of Kenteam; they had to be provided from resources other than the contract. AID did extend help, however, notably in the Darmaga project which undertook to convert a former Dutch estate of 625 acres into a new site for IPB.

With its rubber plantations and cropping areas, Darmaga was assigned to IPB in 1958. The concept of a university campus with planned buildings and functional surroundings only a few kilometers from Bogor entered Indonesian thinking, and counterpart rupiah accruing under U.S. assistance programs were allocated for buildings to house the departments, student ashramas (dormitories), and a few faculty residences. In 1961 President Sukarno rejected the plans and demanded that they be redrawn to provide structures that would "last 100 years—not 25" and would be commensurate with the status and prestige requirements of an Indonesian institution of higher learning. Unfortunately, shortages of cement and steel, labor disputes, "corruption," and mounting inflation were insurmountable obstacles. An agricultural engineering teaching lab was set up as early as 1961, but after six years of intermittent effort only the residences had been completed and occupied, and work was nearly at a standstill for lack of funds. The original plan to move the whole campus to Darmaga was later modified: the basic science work of the first three years would be continued in Bogor; the work of the fourth and fifth years would be conducted at Darmaga. It was not until 1968 that one main building at Darmaga could be occupied by the forestry faculty and some of the dormitory space could be used.

It was, however, a major obligation of Kenteam to assist in the purchase of teaching equipment and books. Ordering commodities was an immediate preoccupation of the first Kenteam members. The purchasing program started immediately and never slackened until uncertain relations between the U.S. and Indonesia resulted in an AID "freeze" early in 1965, later lifted briefly to permit partial fulfillment of plans before Kenteam's departure.

Acquisitions of equipment under the commodity purchasing provisions of the contract included buses to transport students and staff; boats for fisheries "field work" at sea; microscopes; slide rules; electronic equipment; soil testing apparatus; incubators; projectors and other visual aids; supplies of fertilizer, chemicals, glassware, and seeds; distillation equipment; sterilizers; homogenizers; a small saw mill; portable generators, freezers, centrifuges; a very few cars and tractors; autoclaves; air conditioners; a calculator or two; numerous other items, and a few thousand books.

In this context it is not the resulting inventory of supplies and equipment but the strategies of planning and deciding that are important. The contract established a special structure of regulations: orders were placed by the University of Kentucky through its purchasing office on the basis of bids from American vendors and for shipping by American carriers. Orders for any item costing $1,000 or more required prior AID approval. Equipment for IPB was imported duty-free; the Indonesian government paid for in-country transportation and for any costs of installation.

In preparing teaching labs, shops, lecture rooms, and libraries, the process of equipping is a function of the process of education. But making a shopping list in the modern world is easy and temptation takes subtle forms when there are catalogs from which to order and funds to guarantee payment. A problem in the beginning was that planning was behind budgeting instead of ahead, and it never caught up. If there were immediately felt needs, they were the basis for seeking equipment and facilities (commodities). If no felt needs surfaced quickly, there were equipment lists and pictures to suggest needs—too many of them. Pondering commodity needs was an elemental form of programming and planning, and this may have been the context of some of the most effective teaching and counselling that was accomplished by Kenteam members. These side effects of the ordering process may have been as important to the whole project as the acquisition and use of the equipment!

The policy-and-procedure problems in getting equipment illuminate several nooks and crannies in the structure of a technical assistance program.

It was difficult to establish priorities on the basis of teaching needs in disregard of personal interests of IPB and Kenteam staff, and to procure items needed by students and departments rather than in the research laboratories of professors.

It was difficult to stay within limits of acquisition which it would be feasible, in the momentum of development, for Indonesia to continue to support in later years.

It was difficult to meet IPB's every need, and especially to insure future maintenance and acquisition of parts, by restricting purchases to American products. From the standpoint of American interest, of course, the "buy American" policy had a tendency to "keep dollars at home" and to introduce American products to a widening market. It is true also that equipment from America was usually of superior quality. This all helped to stimulate favorable attitudes toward foreign aid by American business and the American Congress. It seems rarely to be understood in the U.S. that most American payments for foreign aid are made to Americans and are spent in the U.S. In the Indonesian view, however, if the acquisition of scientific and agricultural equipment from other sources, say Germany or Japan, had been permitted, it would often have been possible to purchase locally, arrange faster delivery, get better adapted items, and assure easier repair and replacement.

It was difficult to prevent duplication of orders for equipment that could be shared or given common storage or dispersed from common stock rooms or pooled among departments. The disposition to share was no stronger among Indonesians than it usually is among Americans, and only compelling requirements for economy could force its application to the use of equipment and supplies.

It was difficult to enforce the AID requirement that a contract-purchased item be marked with a decal making known to all observers that this item was purchased with American dollars and was a gift to the recipients. Whatever the design of the emblem on the decal, it embarrassed the recipients of aid by exposing their need; they thought it ungentlemanly of Uncle Sam that he would advertise his generosity. Some recipients were merely embarrassed, some were offended, and none cooperated willingly in the marking venture. AID auditors were constantly demanding that Kenteam enforce the decal-identification requirement, and Kenteam was regularly maneuvering to achieve compliance without damage to the sensitivity of colleagues. In a country of brown hands, it was not overlooked that the emblem was a white man's handshake. If the design could have shown a pair of brown hands palm to palm, and another pair of white in the "sila" posture of an oriental greeting, and if the American shield could have stood in equality with the emblem of Indonesia, or at least the merah putih (red and white flag), perhaps the emblem would have been accepted.

It was difficult to withstand pressures to put money into the bottomless pit of transportation needs for cars, station wagons, trucks, and tractors.

It was difficult to specify items of equipment which would operate with available water and electric current delivery systems, which could be housed adequately and safely and operated continuously, with regular cleaning, maintenance, and repair.

It was difficult to assure that workers would be trained to inspect, care for, maintain, and repair equipment and keep up current inventory procedure. It was not always remembered to provide operating and maintenance manuals. The view was never overcome that "anybody with a license can drive." Anticipation of obsolescence and of future needs for parts was difficult, and there came to be cannibalizing and stripping, more in emulation of American junkyard procedure than of institution-building. It was difficult to provide care against mold, insect damage, soilage, rusting, or other deterioration in the humid climate of Bogor. This required tropicalizing, dehumidifying, drying, and air-conditioning.

It was difficult to keep current catalogs of manufacturers and supplies for faculty use in writing specifications.

It was difficult to order against emergency needs, to receive delivered commodities, and to route them to legitimate care-takers and users.

It was difficult and vexing to get delivery in time for Kenteam members to complete installation and training of colleagues in the use of equipment before their end-of-tour phase-out and departure. Kenteam's campus coordinator in Lexington computed ordering and shipping time as about fourteen weeks for items costing $100 or less, but about twenty-two weeks for more expensive items. Kenteam and IPB never advanced the timing of their decision-making enough to beat this delay factor, but they were always more critical of those who purchased, shipped, and delivered than they were of their own tardiness in planning.

It was difficult before even solving first problems to deal with the influence of rising expectations within IPB itself, as years of affiliation passed and especially as participants returned from study in the U.S. Toward the end, the expressed needs of returned participants were given special ear: the wish was to encourage them to introduce better teaching and to begin some research in continuation of their own professional development.

All of the foregoing problems were overcome to some extent, some almost completely. The proof of this was in the final inventory of delivered items and in the record of their use. Difficulties were prevented or overcome by procedures designed to accomplish planning and programming, to facilitate decision-making, and to accomplish ordering, delivery, and use of equipment within the context of the whole Kenteam-IPB relationship.

Concluding Note

The contract provided the law for Kenteam's work but solving each of Kenteam's problems called for special procedures. The body of formal and informal rules and procedures on the one hand drew attention to and strengthened the team as a group, but the real purpose was to facilitate the work of each member, one by one, in his task in a department of the Agricultural University. Some of the ground rules involved the authority of the office of campus coordinator, some involved both home office and field office, but most of them were specific to the work "on post" at Bogor. Some regulated work, some were to serve the needs of the American families in their Bogor homes.

Ground rules were involved in the staffing of Kenteam, in the preparation and orientation of its members, and in their reception by colleagues in Bogor. They mediated the relationship of team members with each other, with colleagues at IPB, and with AID in Djakarta, and they set limits to travel, work, and life. The chief interpreters of the rules were the administrative officer and the chief of party—the latter if reference to policies had to be clarified.

But the real significance of ground rules in an evaluation

of the project is in the clues they give to the concept and spirit of the whole operation and the priorities in its routines of implementation. The ground rules reveal the degrees of confidence and doubt the Americans had in each other and in their Indonesian colleagues. They show the balance of tangibles and intangibles in rewards that Kenteam members expected for living and working so far from their homes. They display the balance of authority and democracy in making decisions and in performing tasks. The ground rules reveal the relative weights put on complete intercommunication, on joint Indonesian-American responsibility, on expenditures for equipment, and on planning and program development.

4. The Development of Educational Programs and Relationships at IPB

A Change in Objectives

Teachers in colleges and universities are sometimes doctrinaire about methods. Most follow the practices their own teachers used and will defend them when challenged. The Dutch professors at Bogor relied almost entirely on lectures, which students entered in their notebooks and memorized. Examinations tested their memory. Kenteam teachers, firm in rejecting the Dutch system, were certain their own teaching techniques were better. These had developed under the aura of American preferences for newness and change, the influence of research in educational psychology, and the urgency to make practical use of knowledge. In simplified form the objective of the Dutch was memorization of factual knowledge; the expressed objective of the Americans was to develop skill in applying knowledge to the solution of problems. So Kenteam brought to Bogor a veritable bagful of teaching methods and techniques. But not many of the American teachers had studied enough of the psychology of learning to prove the merit of the methods they so earnestly espoused.

Time Problems

Organization of its work with reference to the yearly calendar and to daily schedules posed a set of problems never completely answered for IPB. There was a new and revolutionary sense of urgency to educate nation-building personnel as thoroughly and as quickly as possible, but in the background there was also the Indonesian norm of patience and resignation by which man is more the servant than the master of time. At the outset some students were still working for their degree after as long as nine years of part-time or casual academic effort.

Soon after Dean Tojib's assumption of office, decisions were made to revise the curriculum in agriculture and shorten it from five and one-half to five years in length. The Dutch approved but some Americans, in their tradition of a four-year baccalaureate, thought this did not go far enough. Students confronted the dean with resolutions accusing him of being pro-American; he met them and persuaded them to his view. In holding to the five-year Ir. (Insinjur) degree, IPB committees and faculties maintained that it was equivalent to the American master's degree and that they saw no way to compress the curriculum into a shorter program. There was even some feeling for extending the veterinary degree to six years. It was decided, therefore, not to award any intermediate degree after four years, so the program was stabilizing at five years at the time of Kenteam's departure from Bogor.

The five years were arranged as follows: Persiapan (freshman), the preparatory year; Sardjana muda I, or junior scholar I (sophomore); Sardjana muda II, or junior scholar II (junior); Sardjana I, or scholar I (senior); and Sardjana, or scholar (graduate student).

The semester system, however, was adopted in principle soon after Kenteam's arrival, and this prepared the way for increasing the number of courses. In another effort to add new courses, a trial of the quarter system was made in 1962. This system was used first in the agricultural college, but was not uniformly extended to the veterinary college or to all class levels. It was discovered that as much total time is required to cover a given instructional program during a quarter as during a semester, for in a quarter either content must be reduced, or the number of classes per week must be increased, or there must be more hours per class. None of these responses was uniformly made and the experiment ended with a return to the semester system.

Then the length of the semester came under discussion and there was sentiment for a preplanned calendar which would specify dates of beginning and ending, indicate recognized holidays and special events, and fix the timing of examinations. First the faculties, then the whole Agricultural University wrestled with this insistence that calendars be announced well in advance of the college year. Eventually this was achieved and each college had its own Pedoman (catalog) with calendar and courses displayed therein.

Since curricula had to be products of growth and compromise rather than objective standards, and since each teacher was zealous for his subject, it was difficult for a growing faculty of Indonesians with their new American colleagues to reach concensus on the allocation of time to disciplines, to professors, to students—and to extracurricular demands! Differing criteria were bound to produce differing views on choosing four, five, or six years; on setting a semester at fourteen, sixteen, or eighteen weeks; on designating courses to be required of all students; on establishing the length of the class period, the number of class hours per week, the hours when libraries should be open, or the number of allowable class absences.

With their "back home" tradition that time is scarce and must be economized, Kenteam members continued to be perturbed by time problems. Their presence in Indonesia was for a limited period, they were able to start only slowly, and they felt hurried. One typical complaint by an American teacher was over his two classes scheduled as one meeting per week for four hours in the middle of the day, from 10:00 A.M. to 2:00 P.M., when tired and hungry students (whose lectures had begun at 7:00 A.M.) could hardly be expected to pay attention in the heat of a tropical day. The practice of having lengthy class sessions had grown partly from the large percentage of part-time teachers who could be in Bogor only one day a week. This problem declined as the proportion of full-time teachers increased.

It was also difficult for Kenteam members to overlook the numerous interruptions of the schedule by absenteeism, withdrawal of students for ceremonies, holidays, and unscheduled assignments such as road building, military drill, or dispatch to the field for special tasks. Part of the problem was the failure to arrange class schedules so that no student was expected to be in two places at once. One solution was the effort to schedule field trips, tours, and field work in the period between academic years, a move made easier by the even, nonseasonal year.

Toward the end of Kenteam's tour, the credit hour (as distinct from contact hour) and quality-point measures of effectiveness in the use of time were beginning to be understood and accepted at IPB as helpful in evaluating student performance and merit.

Curriculum Development

Curriculum development was a central interest in IPB colleges and departments and there was continuous growth in scope and in number of courses. Most of Kenteam's first members were assigned to teach courses already established in the curriculum. They found that Indonesians didn't "take courses" but "followed lectures," and these phrases are not identical in meaning. The established curriculum was a set of courses-of-lectures and examinations, and most of the lectures were in the sciences, more basic than applied in their reference. However, curriculum modifications were already desired by IPB and early attention was given to improvement of the academic program.

From at least 1959, IPB was continuously in transition as curriculum requirements changed. Students were often puzzled to know their status vis-a-vis the new requirements, which might not yet be in final form. In the meantime, there was the "testimonium," a device inherited from the Dutch for transitional use between the old and new curricula. Kenteam's first wave of members arrived with enthusiasm for applied courses, including even such unorthodox areas as extension education. But a curriculum is under the control of a faculty in any university, old world or new, and innovations must be incorporated by the formal action of that body. A new subject or course could be instituted at Bogor only as a special series of lectures for voluntary attendance, without any required examination and with no credit toward graduation. In the first year of Kenteam's presence it was necessary to accommodate to the presence of some new visiting professors by allowing them to begin under the testimonium arrangement, pending submission of course plans for faculty approval. Curriculum expansion began almost immediately, and the testimonium was rarely used after the first academic year, as the "free-study" system of traditional Indonesia yielded to the "guided-study" systems of free education in . America.

At first all students in each college took all courses ("followed all lectures") in that college. The need for some thrift dictated that the principle of "departmental unity" must be applied as one of the first curriculum improvements in the basic sciences of chemistry and physics, each being taught only in one department but to the students of all colleges. The only specializations were veterinary medicine in the veterinary faculty and agriculture in the agricultural faculty.

There was early and vigorous discussion about animal husbandry, and the presence of Americans fanned the flames. The tradition in the Netherlands and Indonesia had placed all animal science in the veterinary faculties. The tradition in the U.S. was to consider animal husbandry a field of agriculture. In Bogor an Indonesian decision was implemented by stages. First, animal husbandry was recognized as a specialization (direction) within the college of veterinary medicine. The second and final stage occurred when the Agricultural University was established in 1963 and animal husbandry became one of the five charter colleges, being in neither veterinary medicine nor agriculture.

Meanwhile in agriculture a similar cell-division was in process. Forestry became first a "direction" and later another of the five charter colleges, with specialization in forestry beginning in the second year.

The differentiation of instruction in fish sciences was even a little more complicated. Two nuclei formed simultaneously but in different places. A special program in marine fisheries was established in the veterinary college and one in inland fisheries was organized within the agricultural college. When the Agricultural University was formed, these two special units were withdrawn from their parent faculties and combined to form another of the five charter colleges.

In the tortuous planning for these curricula, there were many active agents. The busy "extracurricular" exchanges of opinion among staff members, the views of ministry officials, the advice of deans and department heads, the proceedings in faculty meetings, the work of occasional curriculum committees, the presence of Americans, the assertion of national priorities in national development—all were involved. The decisions were resultants of these forces, compromises acceptable to those involved and having authority. The presence of Americans was surely not the major factor. In the resolution of some issues, their service may have been only in offering suggestions. The suggestions could focus thinking that would lead to the making of decisions. As a national institution, IPB's organization and curriculum were subject to the constraints and prescriptions of the Ministry of Higher Education. However, the ministry and the system of national universities and institutes were younger than IPB, and in the agricultural sciences IPB decisions and recommendations were pioneering for all Indonesia.

In a report by Dean Tojib in 1959, a major revision of the agricultural curriculum was announced, eliminating a division between natural and social science students in the second year, in order to get "one type of generalist" from the lower division curriculum. Fourth-year students would be allowed to choose one of fourteen majors, with two minors, and three electives. The fourteen options were crop production, horticulture, soil science, soil morphology and genesis, soil classification, soil fertility and plant nutrition, entomology, plant pathology, social economics, agricultural policy, agricultural economics, farm and estate management, rural sociology, and agricultural engineering.

Kenteam's curriculum suggestions in Bogor were based on contemporary practice in American colleges of agriculture. By American standards, they were relatively conventional and orthodox rather than dramatically innovative, reflecting

85

established approaches rather than the ferment of change than being experienced on American campuses in response to ongoing scientific, technological, and organizational revolutions in American agriculture. The scientific content of Kenteam's contributions was mainly modern and up to date.

Faculty curriculum committees usually got lost on the job and dissolved informally after a meeting or two without agreeing on plans for reorganization. They did serve as screeners of ideas; their activity stimulated informal discussion between faculty members generally, and administrative decisions were influenced by their views. A departmental curriculum-planning marathon was reported from one group, which held a nine-hour meeting. Intensive curriculum planning followed the establishment of IPB with its five colleges in 1963, and the Kenteam members who worked in fisheries and in the faculty of agricultural technology and mechanization (fatemeta) were especially involved at that time.

Two committees were named in 1963, one for undergraduate and one for postgraduate planning, each with an IPB chairman and an American secretary (one from Kenteam, the other from ADC, the Agricultural Development Council). The fisheries curriculum—not completed until 1965—was a joint effort of the whole fisheries staff with strong leadership from Kenteam. Kenteam on one occasion prepared an outline on curriculum planning principles for use at a national conference of deans.

Postgraduate Work

The IPB staff considered its Ir. degree the equivalent of an American master's degree, which Kenteam was willing to acknowledge in some but not all cases. IPB thinking was that the last two years of its five-year program were postgraduate years. Kenteam, on the other hand, thought of only the fifth year as graduate work, and this view became fairly common toward the end of the Kenteam period. In 1968, two years after Kenteam's departure, this was changed again by a ministry regulation to increase the program to six years, beginning with an orientation or remedial instruction year to make up for deficiencies in the secondary school program. The result was a program similar to the one Kenteam had earlier proposed—a four-year baccalaureate followed by two years of graduate work.

In 1959–1960, when IPB and Kenteam together had reworked the curricula for the first three years, attention turned to the fourth and fifth years, which were classified as graduate years. Interest developed also in arranging doctoral study for Indonesian staff, especially for returned participants who had already completed some advanced work in the U.S. IPB's leaders expressed the hope that Kenteam, increasingly after 1962, would concentrate on helping IPB staff with advanced study in a new emphasis upon an "upgrading" program. As rapidly as returned participants could take over the lower division teaching that Kenteam had been doing, Kenteam members worked in advanced courses and in some low-key introduction of research. In 1964–1965 almost the whole team taught in a committee-devised sequence of seminars on research methodology for IPB staff working toward higher degrees. This effort was clearly moving in the right direction and would have continued in 1965–1966 with an Indonesian staff member in charge, but everything fell apart and the project was not resumed. This was a first step in making additional course work a requirement in Indonesian doctoral programs. Previously, the only requirements after the Ir. degree were the completion of a dissertation, representing original work and acceptable to a professor who served as "promoter," and public examination by the faculty.

Kenteam members were advising individually about fifteen IPB staff members on research intended to help them along toward a doctoral dissertation, but none was finished before Kenteam's departure. In 1965, IPB responded to the national enthusiasm for "standing on one's own feet" and took steps toward emancipation from dependence on foreign study for the Ph.D. by seeking to strengthen its own work. Figuratively looking over his shoulder at Kenteam, a rector once observed that an American with an M.S. degree would be inferior to an Indonesian undergraduate in knowledge of Indonesian agriculture and in use of the Indonesian language.

Approaches and the Parade of Practices

In preparing to teach in Bogor, Kenteam members faced certain general problems of approach and viewpoint as well as procedure. In gaining a concept of IPB's educational mission, program, and facilities, it was necessary always to have regard for the proprietary right of Indonesians to their own goals, their own systems and institutions. It was an ambition of most Kenteam members to make course content less theoretical and to impart to students an understanding of scientific method and habits of scientific approach.

The usual American teaching practices were lecturing, with and without audio-visual aids, laboratory exercise for teaching as distinct from laboratory research by staff members, supplemental reading and library study, classroom discussion, the quiz as a teaching aid as well as a means of grading, the written assignment and the term paper, field work and field trips, seminars for staff and advanced students, student-teacher conferences, student research, special reports on articles from scientific journals, receiving and answering questions from students during class, required attendance, and keeping records of attendance and performance. It was not until the fifth year of Kenteam at Bogor, however, that a faculty seminar on teaching was held, with presentations by Kenteam members, organized especially with consideration of quizzes, testing, and grading.

It came slowly to students that they might interrupt a professor to ask questions, that they might go to a professor's office on their own initiative to seek counsel. A student given an office appointment expected an assignment and was surprised when given freedom to choose a topic on which to write.

A point often overlooked in the advocacy of any teaching method is the skill in its use. A botched field trip is not as good as good lecturing; a well-planned and well-managed field trip is surely better than slipshod lecturing. The quality of the teaching is more important than the technique or the vehicle. Some Kenteam members were more concerned with employing their favorite methods than with trying to be effective in whatever method!

Not everything called a seminar was one. Some were administered rather than held. The professor who reported a seminar in which forty students met for two hours each week was probably incorrect; seminars are not that large! More likely correct was the professor who wrote, "A weekly seminar was held as a regular part of the course, in which each student presented a report." As the program developed, it was somewhat encouraging to see good use made of the seminar technique as a means of creating rapport between staff and students.

But American-brought methods consume time and energy and they require materials. To lecture well requires preparation. If teachers are part-timers or if they "moonlight" to supplement their meager income, they may lack the time necessary to plan a course well and prepare each lecture. Time is similarly required to grade quizzes and for postquiz conferences and remedial work. Textbooks cannot be used if there are none. Library work cannot be done if teachers don't insist on it, or if there are no libraries, or if libraries are not kept open for use. Laboratory practice requires space, equipment, and much more teaching time than lectures. Field trips require preliminary planning and money and facilities. Although Kenteam's effort to introduce new teaching practices was reinforced eventually by the return of Indonesian staff members from universities in the United States, where they had followed American methods, the lack of time, energy, and material remained insurmountable.

Many senior staff members were initially skeptical about American methods at IPB and some became even more confirmed in their doubts as time passed. Under the old system, things were well learned and well remembered; by failure and reexamination, a student was put on his own mettle. If he wished to succeed eventually, he had to memorize the necessary information. If not followed intelligently, the new systems made carelessness possible; students might be advanced without rigorous testing; they might become less self-reliant, more dependent on the teacher-with-the-new-tricks. The real reasons why teaching did not change faster and permanently, however, lay not in the lack of merit of the new methods but in the perennial shortage of time, money, and materials. The practice of paying the teacher a fee for each oral examination or reexamination given may also have inhibited the adoption of new practices.

While a Kenteam member was physically present, the teaching methods he preferred were more likely to be used than after his departure, partly because his time and energy cost the Indonesians nothing and partly because of his access to materials that his Indonesian successor would probably not be able to acquire. The most lasting impact was made in those cases where a Kenteam member and one or more Indonesian colleagues worked and taught together, using adequate books, teaching, and equipment, and where later an American-trained Indonesian returned ready to carry on the new practices.

Adequate teaching preparation was a major problem facing Kenteam teachers and their associates. After Kenteam's third year, one member commended his chief Indonesian associate, recently returned from foreign study, for "an excellent job in teaching. . . . To my knowledge, she is one of the few instructors who has made up a complete outline of both her lectures and her labs and handed them out to the students before the course even started." Most, but not all, Kenteam members who taught classes did follow such a procedure. One report, for example, is that "My first task on returning from home leave was to spend a solid month writing fifteen lesson plans for lecture and laboratory for the new course." But no indication is given that any Indonesian associate was jointly involved in this effort. Gains were made on this front, too, however, and it finally came to be a requirement that outlines of all courses be filed in the offices of the deans.

The need for textbooks in Indonesian and with Indonesian application was often cited: in livestock production, poultry, herbicides, agricultural marketing, economics, and actually in nearly all fields. In lieu of books in Indonesian, books in

English were provided for many courses from contract funds. Some Kenteam members became involved with their colleagues in textbook projects, notably in microbiology, plant physiology, and agricultural economics, but textbook writing was not a major accomplishment of Kenteam. The void remained unfilled.

Lectures in English were little understood by IPB students in the early years, although this was acknowledged only hesitantly by Indonesians because of their intense desire to acquire competence in English. One student confessed he "watched the blackboard." Answering examination questions in English prevented the student from revealing his full knowledge of the subject, yet most Kenteam professors could read the examinations only in English. Indeed, some examination papers were valuable mainly as practice in writing English. While it was recognized that writing narrative answers to "subjective" questions requires a student to use information rather than merely to report from memory, it was also true that he could do better in Indonesian than in English.

Teaching with—not without—Indonesian associates was difficult and sometimes impossible to arrange. Even with Indonesian associates available it was not easy for a Kenteam member to be effective. In their annual and terminal reports, all too many Kenteam members told of the courses they taught, the number of students enrolled, and the methods they used, but with no reference to the participation of Indonesian colleagues. In his comment on working with an Indonesian associate, one Kenteam member wrote, "The course was given with the department head as collaborator. This was a pleasant experience for me. The students prepared good reports and made effective classroom presentations. A set of mimeographed lecture material was prepared for distribution to students and staff members." For another course, the associate "was unable to attend lectures regularly and as a result KCT-IPB cooperation was not very successful. The lasting benefits of the teaching of this course were considerably handicapped." One Kenteam member who prepared a complete set of lecture notes in English went over each lecture with an Indonesian associate who did the actual lecturing in Indonesian, felt that he had made a useful contribution, but was somewhat frustrated by his inability to know what lecture content was really presented.

There were some situations in which Indonesian associates did not participate intellectually in the teaching process but performed routine tasks as assistants. It may have been true in some such cases that keeping class rolls and recording grades were useful activities, but actual participation in planning course outlines and lectures and in arranging for laboratory classes would have been more effective as preparation for teaching.

It was the view of several Kenteam members that "one should not expect or require students to follow regular full-speed lectures on technical topics in a foreign language provided suitable alternatives are available . . . that visiting staff members who are unable or unwilling to present lectures competently in Bahasa Indonesia contribute far more by acting as consultants to regular staff members in planning course content and coaching in teaching methods."

Another comment confirmed this view and made an additional point: "After having taught a part of two different courses I feel that my teaching should be confined to helping someone prepare material that will be presented by an Indonesian staff member. If possible any written material to be duplicated should be in Indonesian. It is interesting to prepare and teach any courses in forestry to these students who are friendly, cooperative, attentive, and anxious to learn. The presentation in English, spoken slowly and distinctly, with many explanations of words, certainly helps to improve the students' command of this foreign language. It is highly desirable that a counterpart be assigned to the adviser in whatever courses it is deemed appropriate for him to teach."

The questionnaire, a favorite American device for self- and course-evaluation which was applied in a few cases, proved a novel experience for Indonesians. Three months after a group of second-year students completed their biochemistry, a questionnaire was presented by the Kenteam teacher. They all thought biochemistry "important"; 83 percent found it not too difficult; they studied less than three and one-half hours per week; they thought the quizzes fair; they unanimously approved laboratory experiments and said they liked laboratory work; 80 percent said conferences with the professor were helpful; but the average amount of lecture material presented in English that was understood was only 50 percent. A similar procedure was followed by a Kenteam agricultural engineer who reported, "The main suggestion was to include more practical work and more learn-by-doing activity in taking apart and putting together old engines and working with farm machinery." This evaluative procedure was not adopted by IPB staff members. The traditional desire of the Indonesians to tell the interrogator what they think he wishes to hear perhaps affects the validity of this device.

Kenteam, exuding American pragmatism, seized on the IPB "prakticum" as an expression of the caveat, "Learn by doing." The teaching lab, the clinic, the field lab, field trips, field work and "interning," papers, individual work on problems, and participation in surveys and other research projects were all methods used by Kenteam members in their courses and counseling. Students in horticulture visited horticultural experiment stations which could be reached within two hours of travel from Bogor, Pasar Minggu and Tjipanas. Plant breeding students visited the Cereal Crops Research Institute and students in some courses, including marketing and farm management, made trips to other research institutes in Bogor. Longer tours took fourth-year veterinary students to Java, Madura, and Bali. A marketing study tour for staff members (not students) touched marketing centers in the three provinces of Java. A simple study of egg quality on the Bogor market involved Kenteam-IPB collaboration in two departments. An ambition of Kenteam members to demonstrate what good broiler chicks would do in contrast to the meat of scrawny village Java fowl, however, was not realized while Kenteam was in Indonesia.

In forestry also there was an emphasis on field trips, tours, and field work in special forest locations. A demonstration-research forest and camp were maintained at Pasirmadang, several hours from Bogor, but there were chronic problems of transport and budget. Extensive field work involved going to a location and conducting some study, survey, service, or research project.

IPB's work in both inland and marine fisheries, for example, was heavily weighted with laboratory and "prakticum" experience, on and off campus. It was a dream of the marine biology faculty to organize an annual or biennial research and training cruise of two or more months to be undertaken in cooperation with the navy, but this had not yet been achieved when Kenteam left Bogor. On one occasion two students were sent to Kisaran in Sumatra to work in the field on the extensive lands of an oil company with some guidance from the technical staff there. Field work was not feasible in all years or in all courses. Some classes became too large and there were persisting problems of limited transport and equipment and conflicts with other classes and laboratory periods. Vacation work was soon expanded to all provinces for both agricultural and home economics students, a forerunner, actually, of the Bimas scheme which attracted so much attention in 1965 and thereafter as a means of increasing rice production. It began with summer assignments to work in the extension services or on estates, made by a Kenteam professor of extension methods.

Surveys were a preference of some Kenteam members, but they presented special problems. The formality of survey procedure, with structured questions and uniform character, were foreign to the habits of inquiry and intercommunication in villages. It had already become a part of the curriculum before the days of Kenteam for students in the second year to go for two weeks to observe village life, and again in the third year for two weeks to study the operation of farms. Second-year village visits were part of the developing program of instruction in rural sociology; third-year village visits were part of the developing program in farm management. A major objective was to give all students—of whom only a few knew about villages—at least a minimum knowledge of agriculture and rural life. The purposes of summer work after the third year, of field problems for students in the fourth year, and of research in the fifth year were more directly related to the curriculum. In fisheries, it came to be the practice that all fourth-year students would go to the field for practical experience during the second semester. At this time each would carry out a short research project or practical field work assignment in his major and minor fields. Project outlines for the field work were prepared and approved in advance, and final reports of the work also were submitted and approved by the staff member making the assignment. There was general acceptance of this program throughout the Kenteam era, but many criticisms as well. Kenteam was restive about the annual two-week interruption in scheduled class work for which additional time was not provided and some wanted the scheme abolished for this reason. Moreover, it was felt that the field experience was too casual, should require more careful preparation, more systematic activity in the field, more field supervision, and better on-campus follow-up and integration into on-campus course work. IPB staff in economics and sociology made some improvement in field procedure through prefield orientation lectures and tests; instruction in group interviewing; clarification of definitions; arrangement of lists of questions with a thought to later editing, coding, and summarizing; and preparation of manuals of field procedure. Instruction in research methodology was introduced by the younger staff members, several of whom had had recent training abroad.

Not only was IPB using this improved procedure, but it was beginning to yield research data. Plans to reproduce reports of field work, however, were retarded, and in some cases abandoned, because of paper shortage, intervention of other duties, and a variety of other factors. One innovation was the course in rural sociology, conducted after completion of field work, in which small groups of students would present, by group discussion procedure, their report of village experience.

Sudden enlargement of the prakticum idea in 1964 took Kenteam somewhat by surprise. All fourth- and fifth-year students were scattered to the provinces to assist in the program identified bilingually as SSBM (self-sufficiency in bahan makanan, food materials). The purpose was to encourage farmers to fulfill the pantjausaha (five practices):

95

using good seed, planting in rows, fertilizing, weeding, and insect control. To the perplexity of Kenteam, this program meant that the student suddenly disappeared! Many doubted that immature students without village background and with meager training could be effective in this work. This program was later changed to Bimas (from *bi*mbingan *ma*ssa, mass guidance), and it is enough to say here that production did increase where the students worked and that Indonesia had discovered a new method of extension work.[1] Bimas, which involved sending students to the villages, was extended and expanded in subsequent years, but seemed to lose some of the effectiveness that the spontaneity and dedication of students were able to produce in the first two or three years.

Some effort went into obtaining background material and information for courses of instruction. In home economics, for example, many class periods and conferences were occupied by study of similar projects in Puerto Rico, the Philippines, and Taiwan. The result was a detailed survey plan which two students carried out in two different villages, living in the areas for two months and studying village life as it related to water supply, bathing, disposal of waste, drainage, kitchen activities, food storage and preparation, and housing.

Writing "scripsis," or topical papers, was an established practice at IPB, but Kenteam members thought some aspects of this were overdone, others underdone. Some felt that the three papers required did not permit adequate reading and research, and that consequently the student was unable to go deeply into any subject.

The provision of teaching laboratories was an early and continuous interest of Kenteam at IPB. Priority was given to the laboratories and stockrooms for chemistry, physics, and other science courses; this was a major expenditure under the UK-AID contract. Funds were reserved mainly for teaching-laboratory and not for research-laboratory equipment (only

[1] D. H. Penny, "Agricultural Extension for the Masses," *Bulletin of Indonesian Economic Studies* (Canberra: Australian National University), no. 2 (September 1965):60-63. Leon A. Mears and Saleh Afiff, "A New Look at the Bimas Program and Rice Production," *Bulletin of Indonesian Economic Studies*, no. 10 (June 1968):29-47.

in the later years could some attention be given to research needs as such), and continuous efforts were made to integrate the laboratory and lecture sequences. A soils laboratory and a plant physiology laboratory were installed, and the Kenteam member in charge introduced a one-semester laboratory course dealing with measurement and analysis, especially mineral analysis of micro and macro elements. In the animal sciences large classes were divided into lab groups of five or six students. The problem of obtaining clinical materials, particularly laboratory animals, was difficult because of lack of money.

In food technology a professor wrote in 1963 that there was "a fairly well-equipped but overcrowded laboratory with students swarming within a few feet of my desk much of the time." He reported, however, that this informal action-contact was a great advantage over formal written and verbal consultations.

Field laboratory work was introduced in several courses. In poultry, for example, it was reported that Indonesian staff members did a "splendid job" with a student chick-raising project. One of the most interesting examples of field laboratory procedure was in horticulture. Land beside the horticulture building was divided into seventy individual plots, each two by three meters in size. Each student was assigned a plot at the beginning of the term and all were required to plant the same crops (tomatoes, cucumbers, beans, and sweet potatoes) and employ the routine cultural practices, including planting, weeding, watering, fertilizing, and pest control. In the first class in agricultural engineering offered by the faculty, machinery was assembled, serviced, and put to work by students. In forestry, a field laboratory was established and instruction was given in the use and maintenance of hand logging tools.

Examinations

Changes in the old examination system came about because of the vacuum of technical and professional manpower and Indonesia's haste to produce graduates, rather than just through the efforts of Kenteam. Kenteam members were startled to learn that failure rates ran as high as 80 percent under the old system, although most students who failed would pass later reexamination. Eight or nine years often passed before graduation. The new "guided study," as it came to be called by Indonesians, provided at least a partial solution. Along with insistence on required attendance came a change in examination procedure. Occasional quizzes were held, some announced, some unannounced. Make-ups for absence were not allowed; some quizzes were objective and some subjective in format. Grades, previously withheld, were now made known to students in order to make them aware of their own rate of progress. Classroom discussions occurred; professors explained the reasons for quiz failure and held individual conferences with students.

The mechanical problems of administering a quiz in a crowded classroom, with rows of students side by side, made cribbing a problem. But by norms of mutual aid—the customs of equally sharing the limited good—helping a fellow student to answer a question was thought to be a social requirement and was not considered cheating. By the values of the villager, this was "social justice" and not at all un-ethical. On at least one occasion an angry Kenteam member was much disgusted with his chief of party for cautioning against making an issue of cheating, proposing diversion through a new method of examining rather than through confrontation for crime or castigation for sin. A device introduced by one Kenteam physics teacher was the one-question quiz, with a different problem for every person in a row.

The old ten-point grading system[2] was retained for official records, but Kenteam teachers used their own preferred A/B/C or percentage records. If a student's final grade meant failure in a course, Kenteam members sought to require that the course be repeated before reexamination could occur. In the determination of a final grade, Kenteam sought to establish

[2] One and two, very bad; three, bad; four, insufficient; five, next to sufficient; six, sufficient; seven, more than sufficient; eight, good; nine, very good; ten, perfect. Six was the minimal mark in order to pass the examination.

criteria other than the traditional professional evaluation of performance in an oral examination. They used quiz grades, attendance records, laboratory reports, and papers in addition to the mark on a written final examination. One professor—probably typical—wrote that "laboratory work and reports count one-third, quizzes count one-third, and final exams one-third of the final grade." One Kenteam teacher wrote, "In the 1958-1959 year a total of 140 students took the course in Plant Breeding, and approximately 83 percent passed."

Failure was rationalized differently. To the Dutch professor under the old system it was student failure. The student had failed to learn. To the newcomer on Kenteam, it was instructional failure. The course had not provided true "learning situations" and the professor had failed to teach. No doubt both views are extreme and the blame should be shared, but the need was for a system geared more to the students than to the professors. It is probably good for a professor to tell himself that all student failure is his failure, a view widely held in American education, not because it is entirely true but because it induces him to strive to improve his methods.

Certain controls over grading were reserved by examination committees; this puzzled and even irritated Kenteam mildly. The custom for American teachers to grade their own students is strongly established. To have the grading done by others who know neither the professor nor the students and have not seen the teaching situation is upsetting, to say the least, although calm logic may note the possibility of some objectivity in such a procedure. It is clear that external examination is completely divorced from counseling, however, and if evaluation is construed as an active feature in education, it must really be a joint undertaking of student and teacher, and open to their continuing and mutual review.

Research

In modern education—even in elementary and secondary schooling, but certainly at higher levels—research has become not only a quest for new knowledge, but also a major instrument of teaching. The ability of a college or university

99

to promote postgraduate study depends upon its resources for research. In the division of labor among agencies and institutions in Indonesia, however, research was the special function of separately established institutes, and faculties of the Agricultural University were charged mainly with teaching. IPB's budget was committed to other purposes and there were no contract funds reserved for research, so it was only incidental to the main job.

Research was beginning to come into view at IPB, however, as a necessary and inevitable method of learning and thus of instructing. Appearing first in some of its least sophisticated forms, research was the exploration of some question in a newly-equipped teaching laboratory, or the preparation of a scripsi reporting special observations, but usually not going beyond descriptive exposition. The scripsi, an inheritance from the Dutch, was an established part of IPB's educational program before Kenteam arrived in Bogor.

Most American agriculturists are so accustomed to living in the presence of research that a typical member of any group like Kenteam will either initiate one or more projects or suffer feelings of guilt. Having accepted and practiced an advisory role in Bogor to encourage the initiation of research, to provide some help in supplies and transportation, to be available for consultation and discussion, and in some cases to promote cooperative work, one team member reached the conclusion from his experience that "each team member should strive to conduct some suitable research of his own with Indonesian assistance. Merely helping and advising someone else in his research is not an effective training program." Yet the Kenteam member in Bogor was more of an advisory bystander than a real director of or participant in a project. This was part of the philosophy of acknowledging the prerogative of Indonesians to set their own goals, establish their own institution, acquire the knowledge and skill to do their own teaching, and conduct their own research.

Although it was suggested to Kenteam from time to time that some IPB research be financed from American Food for Peace (Public Law 480) funds, this was found not to be feasible because of the difficulty of organizing an approvable project within a time period that would end before the American "principal investigator's" tour in Bogor expired.

Wide-ranging research interests among their ɪᴘʙ colleagues were mentioned in annual and terminal reports of Kenteam members. Dozens of topics came to attention in the various subject-matter areas during the years of affiliation, each a project on which some Kenteam member and some ɪᴘʙ staff member had worked together, at least casually. Many were the topics of scripsis, thesis problems of fifth-year students. An innovation in which teaching and research were formally joined was the organization of projects engaging several fifth-year students to do field work. This gave them experience in problem identification and in the formulation and systematic testing of hypotheses.

Many of the study topics reported work of a fact-finding service performed at the request of some agency or ministry and from which it would be too much to expect scientific generalization. These were, however, means by which various problems could be delineated for stouter treatment if resources should ever become available. It was judged likely, by both ɪᴘʙ and Kenteam members, that for some time to come many of the problems of the Indonesian economy could best be approached by some of the simpler methods of research. A bureau of statistics was formed at ɪᴘʙ in 1965, however, and staffed by well trained returnees from foreign study.

Some of the research interests listed were scripsi topics, resulting normally in manuscripts filed with the departments concerned but rarely resulting in publication. Some topics were still hardly more than statements of intention for which research designs remained to be formulated or for which data were not yet in hand. At one point the coordinator of ɪᴘʙ's forestry research reported that ten papers about these projects "are ready for publication, but that the shortage of paper and printing facilities has delayed this." Taken together, however, the topics confirm the presence of a reservoir of research interest at ɪᴘʙ, filling rapidly with the ideas of participants fresh from advanced training abroad and with new concepts of research purposes, procedures, and possible accomplishments.

Horticulturists did variety tests of seeds: cucumber, cabbage, tomato, onions, peppers, "as a focal point for student research projects, as well as a means of demonstrating the techniques in doing applied field research." Kenteam and IPB colleagues studied the local egg market and started some feeding studies, but research involving large animals would have to await more funds. A virology laboratory became the scene of research with which the Indonesian director gained his doctor's degree. IPB staff undertook to study "cultural techniques in pineapple," "nodule bacteria strain relationship in grassland combinations in the tropics," "predicting corn yield by other plant characteristics," "comparison of two methods of corn breeding," "demonstrations comparing autoclaved and oven-cooked soy beans," "feed supplements for pullets," "cooling eggs by wet cloth," "marketing of vegetables," "the government rice purchasing," "production, consumption, and marketing factors in relation to proportions of paddy and rice marketed," "rice marketing," "credit," "costs of production," "effects of different regimes of photoperiods on some varieties of soy beans," "mineral content of plant samples from Kalimantan," "effect of nutrient solutions on stem cuttings of sweet potato," "the influence of growth substances on the vegetative propagation of tea," "the effect of plant regulators on the rooting of peanut cuttings," "strain evaluation of legumes," "development of Newcastle disease vaccine." The list continued in this fashion would extend to several pages and for the purposes of this discussion could only reinforce the general point being made.

An important incentive for research, it must be noted, was the possibility of an additional wage. For each official job there was payment, and each lecture in another course, each examination conducted, each observation field trip made, each research assignment brought a little extra income. Seventeen IPB faculty members received rupiah research funds from the ministry in December 1962 in a move that caught Kenteam uninformed. They were generally delighted at this stimulation of research, but their invitation to counsel with the awardees and help them to use the research money came a bit late. The IPB recipients had already chosen their problems and stated their procedures, apparently with some prodding from the deans and the ministers of education and research themselves. It was after the fact of the awards that Kenteam members were asked to help. They did, indeed, proffer their services, but some of their suggestions were taken as criticism. Some of the seventeen projects were completed, some were done partially, and some not at all. The rupiah allocated by the ministry were a means of starting some studies, but even more conspicuously a way of supplementing the meager income of those who took part. When IPB salaries become adequate, research activity may become more an expression of scientific interest and less an expression of economic need.

Capacity for Further Growth in Postgraduate Work and Research

When asked to judge IPB toward the end of the affiliation, a minority of Kenteam (41%)—but with many abstentions—and a much larger representation of IPB (76%) agreed, each thinking of his own field, that IPB had achieved a capacity for self-sustaining growth in teaching and research at the graduate level.

The division of opinion over capacity for graduate research is similar to that over graduate teaching. No Kenteam comments expressed full agreement, but a few IPB staff members called attention to the presence in Bogor of research stations, the role of the Bogor community as a "center of sciences," the acquisition in recent years of facilities, and the fact that some faculty had advanced degrees.

An affirmative Kenteam view was that "The research attitude is getting a little better implanted; knowledge of research techniques is a little more adequate; but there is still too much of a tendency for a rather unsophisticated and generalized approach in which hypotheses are not explicitly stated and observations are not sufficiently guided by the canons of research." Only 30 percent of Kenteam members felt that IPB was now ready, at the moment of Kenteam's departure, to provide adequate preparation for doctoral degrees in certain programs. The self-confidence of IPB is seen in their 76 percent agreement.

The following fields of study were mentioned by both Kenteam members and IPB staff members as having Ph.D. capability (but none was mentioned by more than three persons): forestry, agriculture, veterinary medicine, soil sciences, botany, animal science, agronomy, social economics, and crop science.

Hence, with respect to personnel—in number, competence, and motives—IPB by 1963 was assuming a suitable research posture. If there had been a little more financial support and a little more support in the sociopolitical climate, members of the IPB staff would have launched a substantial and useful research program. They had acquired enough expertise and confidence through advanced training overseas to get started; they were choosing topics, problems of importance to Indonesian development; they contemplated methods and techniques that suited a mixture of research and action; they sought professional advancement, and they sought to serve their country's development. Before them lay the opportunity to establish appropriate patterns of research specialization and cooperation with the institutes in Bogor. While Kenteam was at hand, they had recourse to advice and counsel. Other things being equal these factors could have confirmed IPB's place in the sun as a center of excellence, capable of graduate instruction and research, but the climate of the time did not support their mood.

IPB and Other Faculties

IPB's network of external relations included other faculties, Bogor's other institutes and agencies, and governmental agencies and projects, and was mediated through a developing system of extension programs and public service. All national institutes and universities were a system, organized under statutory authority and administered in the Ministry of Education. There were natural rivalries between the institutions, of course. The University of Indonesia in Djakarta and the Gadjah Mada University in Djogjakarta, for example, were competitors for support, students, and prestige. All institutions were in competition for such largesse as the national budget might afford.

One of the most important relationships involved educating cadres to become faculty members at the other colleges, sending ipb staff as "flying professors" for part-time teaching, conducting study conferences and short-courses for "upgrading" the local staff, surveying facilities, and offering counsel. These needs were beginning to appear as early as 1960. In fact, before the relationship between Indonesia and the U.S. reached the 1964 and 1965 low points, the Ministry of Higher Education had planned to ask Kenteam's help in developing a national plan for higher education in the agricultural and engineering sciences. The problem that ipb faced as a mother institution (fakultas induk) in accordance with the ministry's assignment was how to maintain its own strength and even to progress while at the same time diverting energies to the service of other faculties.

As institutions recovered their balance following the difficulties of 1965 and 1967, the designation of helping institutions (pembina) was reaffirmed by the Ministry of Education, and ipb began to intensify its work in teaching cadres for other faculties, in sharing its resources, and in advising with other faculties on all features related to their work.

IPB and Bogor's Other Institutes and Academies

Scattered widely in the Bogor community, each with its own facilities, were twenty-six or more institutes for research and academies with three-year training programs. A few members of the ipb staff had part-time employment in other units. Each year certain graduates finishing their work at ipb joined the research or teaching staffs of the institutes and academies; relationships thus were informal, in the sharing of personnel rather than in coordination of planning or activity. Some Kenteam members had the impression that relationships between ipb and the institutes were too competitive—or at least unrelated—and that there was no encouragement from higher up for joint work. The number of joint interests was growing, however, and the day might be foreseen when the research institutes and IPB would be mutually supportive and cooperating units in a coordinated network in the agricultural sciences and technologies. By complex mutual arrangements, some of IPB's students were assigned to the research institutes for their fourth-year prakticum, and there could be offered a long list of persons working at both IPB and an institute, but the institutes had numerous projects in which IPB was not involved, as in soil survey and mapping. Each year during Kenteam's presence, IPB's relationships with the institutes expanded a little and diversified, partly in response to Kenteam's curiosity and uninhibited acquaintance with institute directors and their colleagues. But IPB and the institutes in Bogor were still mainly underdeveloped resources, each of the other. The potentialities of cooperation and coordination among them remained as resources for future development.

Developmental Service

Information and wisdom are needed so desperately in a developing nation that men who feel deficient in both are burdened with responsibilities beyond their capabilities. Many of them turn to agencies and institutes of knowledge and learning for help in making decisions and solving problems. Student soldiers of the struggle for independence who came from battle into the lecture hall, some of them quickly becoming instructors, were experienced in action, and faculties readily became involved in various assignments calling for developmental services to government agencies and projects. A member of Kenteam noted this function of IPB and voiced a warning that it might be "overdone" in the early stages of growth, while the faculties were weak.

The national ministries frequently call upon the faculty to undertake projects of interest to the government or request assistance from university personnel in carrying forward projects for which the government is directly responsible. Members of the faculty serve in an advisory and consultative capacity to the ministries in Djakarta, and an effort is made to direct the training research, the so-called practikum of students, along lines which will provide socially useful research results. All this should be continued and even expanded, but the government must be on guard against too much infringement upon time sorely needed for advanced study either at Bogor or overseas. There is a critical need for trained people in government and business, and at the same time some scientists are needed to expand the teaching program or perhaps should be given time to improve their own proficiency through more advanced training. The choice of whether to use scientifically trained personnel as items of consumption, as a form of capital expenditure to produce more trained people, or as capital goods whose productivity can be increased by further investment is indeed difficult.

To become overinvolved in action threatens reflection and study and eventually may weaken a scholar's or an institution's intellectual capability. But to be underinvolved in action fosters unreality and intellectual sterility. Each scholar and each school must find the proper balance, the proper tension between involvement and aloofness, so that knowledge expands and service deepens. IPB was seeking to serve simultaneously its own growth and the needs of other developmental agencies, organizations, and institutions. The spraying of caterpillar-infested coconut trees was accomplished with airforce helicopters. A survey of vegetable marketing in Djakarta was conducted for the Agricultural Service of West Java. There was a trouble-shooting study of erratically colored water lilies in a presidential palace pond. Cooperation was extended to an Indonesian poultry congress in Djakarta. Major assistance was given in the succession of national programs to stimulate food production, especially rice yields. IPB helped organize and stage the national census of agriculture. It was involved in the analysis of problems and formulation of policies for land reform, for programs that sponsored migration from Java to other islands, and in the organization of national efforts to provide agricultural credit and to stimulate cooperative organization for economic objectives.

Other relationships evolved with officials of the city (the mayor served as chairman of the Board of Curators); with the People's Agriculture, Forest and Livestock Services, for whom in-service training and advisory functions were performed; with the district and provincial governments, with which memoranda of agreement were drawn up to provide for research and service functions; with farmers' cooperatives through in-service training for their personnel; with the farmers' and fishermen's banks, by memoranda of agreement; with ministries other than education, for whom research was done and to whom advice was proffered; with the several research institutes in Bogor. A rubber company in Sumatra was host to a few advanced IPB students, offering them practical experience and technical guidance. Business firms contributed visual aid materials. An international company sponsored a short course in the use of power chain saws. Scholarship awards were made by the American Men's Association in Djakarta in 1961, 1962, and 1963. IPB was becoming also an important sponsor of and participant in national and international conferences and seminars.

Extension and Public Service

The operation of a field service in the extension of agricultural knowledge was vested nationally in a department of the Ministry of Agriculture. The minimum service of IPB to agricultural extension was in the training of graduates to be employed in extension. Extension field personnel, however, had come previously from agricultural high schools and from the three-year academies. Not all agreed that the five-year training for the Ir. degree, or its equivalent, was either necessary or suitable for an extension worker.

The question of how to become involved in extension work attracted the early interest of IPB, especially its first Indonesian dean, who had a dynamic conception of the duties of his office and who took appropriate educational and administrative initiative. This function became a major official obligation of all national universities when the Ministry of Education announced that service to the community must join research and teaching to fulfill the threefold responsibility of a university.

Afterthoughts

The story of IPB's growth as an institution of teaching and learning between 1957 and 1966 is a story of innovation. The process was not alone de-Europeanization, Americanization,

108

or Indonesianization—although all were involved—but it was mainly the last of the three processes. There were innovations in objective and emphasis, admission practice, calendar, organization of courses and curricula, and teaching-learning techniques in classroom, laboratory, and field. The whole effort was intensified; many new practices were teacher-intensive, others were student-intensive in their requirements of energy and effort. Teachers increased their preparation, consultation, and evaluation inputs. Students increased their homework and recourse to books in the library, and accelerated their time schedules. There was a general intention to put more substance in forms, as in changing casual field observation into systematic field research. It was and remains a tragic blow to further development that neither teachers nor students have income enough to be able to give all their energies to their jobs at IPB.

An American visitor might conclude that IPB was transplanting the land-grant college system to Indonesia, but this was not at all the case. The Indonesian leaders and their colleagues knew very well about the land-grant system—their first dean of agriculture had made a study of it on location in the United States, and before the affiliation had ended, nearly 200 staff members of IPB were alumni of graduate schools in land-grant universities. With Indonesian acuity, the staff at IPB was maintaining some of the system they had learned from the Dutch, studying systems elsewhere in Europe and in Asia, refining their own expectations, and utilizing the innovative judgments of Kenteam colleagues; out of all this it was finding identity as the Indonesian Agricultural University. Its multiple function—research, teaching, and community service—fulfilled a very modern concept in highly developed countries, but its exact model does not exist in any other country, nor has it yet settled finally on the shape it will seek to achieve in the future.

All the innovations at IPB resulting from Kenteam influence are now under challenge in the American university, where students want relevance, self-determination, self-effort, participation, real-life involvement, and access to the professor as a resource person and fellow student rather than as a dispenser of facts and data. Paradoxical is the fact that Indonesian students came into the classroom from revolutionary struggle, not from affluence, and continued their political interest and activity while Americans worked among them to innovate philosophies and methods of education. In the same years, affluent students in American universities rose up before their teachers and demanded adulthood and community involvement. An unanticipated consequence of work in a developing country was the reverse technical assistance to American professors, who returned to their home campus better able than before to understand student thought and action.

5. Developing an Indonesian Faculty: The Participant Program

Kenteam's mission in Bogor was to achieve the development of an Indonesian Agricultural University with its own professional personnel in whom would be vested the full responsibility for IPB's broad, deep, and self-regenerating educational performance. All of Kenteam's activities were directed toward this end and probably the most important of all strategies employed was that of providing advanced education in American universities for faculty members, including especially new recruits as they completed their undergraduate preparation and were selected to hold faculty positions after periods of planned foreign study. All parties to the Bogor project in technical cooperation called these candidates "participants." Many Indonesians travelled and studied in the U.S. under the terms of other AID-university programs and many participants were sponsored by AID programs additional to the university contract projects. In the total AID participant program, the 4,000 mark was passed in 1965; it had been planned to celebrate this in a publicized ceremony at Bogor, because participant number 4,000 was a member of the IPB staff, but the political situation at that moment required low visibility and the event had to pass unnoticed.[1] Within the Kenteam-IPB program itself, 204 participants were sent for advanced study in the U.S. before Kenteam's withdrawal from Bogor on February 28, 1966.

IPB participants studied at thirty-five different institutions in twenty-nine states, although more (70%) worked at the University of Kentucky than elsewhere. Just 50 percent of IPB's participants received master's level degrees, and 5 percent received American doctoral degrees. Forty-three percent stayed in the U.S. for one year or less; 48 percent remained between one and two years; 9 percent stayed longer than two years for special Ph.D. programs. Among those who did not receive U.S. degrees, most were judged to have the requisite abilities for post-graduate study but their program did not include a degree objective and the approved time allowance was too short for completion of degree work. Ph.D. degrees were received from ten American universities. Among the returnees who had not aspired to doctorates in the U.S., five had, by June 1968, been awarded doctoral degrees in Indonesia, according to the regulations established there.

In terms of cost the participant program was the most expensive part of the total program, to both Indonesian and American budgets. Interest in the program was keen, with a good deal of support and appreciation on both sides, as well as some controversy, disapproval, and a variety of before and after judgments about the contribution of the participant program to the development of IPB. American graduate study for Indonesians was then a new thing, but now it has been tried—on a rather large scale. College men and women from the Spice Islands of the East have lived as students in Western communities; they have eaten Western food, followed Western curricula, read Western books, and brought back Western degrees.

Features of the participant program included the selection procedure, preparation for travel abroad, provisions for study in America, the students' problems and performance there, and the resumption of their work in Bogor. Three surveys— one of returned participants of whom sixty-eight (39%) responded, one of Kenteam members of whom thirty-eight (81%) from a possible total of forty-seven responded, and another of the American university advisers of participants of whom 60 percent responded—have provided data on the

[1] Characteristics of programs for the advanced study of foreigners in the U.S. and the problems of participants in those programs have been the subject of several reports. For example, see Albert E. Gollin, *Education for National Development: Effects of U.S. Technical Training Programs* (New York: Praeger, 1970). In addition to reporting a worldwide survey of participant training, this book contains a bibliography identifying the main other studies which have been made.

participant program. All three surveys were made in the last three months of 1965.

Selection of Participants

The choice of candidates for overseas study was guided by broad policies in which IPB and Kenteam concurred and which required that candidates must be bona fide appointees to the IPB staff, qualified in English, recipients of the Ir. (Insinjur) or Drs. (Doctorandus) degree, grantees of official leave (with pay diverted to their dependents), and under commitment to return for service at IPB.

The possibility of becoming a participant attracted graduating (Insinjur) engineers to apply for appointment to the faculty, since the experience of the trip abroad was highly coveted. Supporting funds set limits on the number who could go each year, however, and careful selection was necessary, but there were competing views about how staff members should be selected.

It was sometimes said by Kenteam members, generally in criticism of the practice, that the IPB's application of the doctrine of the limited good—expressed as equal treatment for all—resulted in selection of some participants less qualified than others because of the feeling that all fields and departments should have an equal chance without regard to possible differential needs. A majority of Kenteam (53%) and a larger majority of IPB (63%) indicated their belief that the "equal treatment" principle had influenced the selection of participants. A minority (32% for Kenteam; 40% for IPB) felt that this principle was involved in decisions about whether the foreign study tours of participants should be extended to allow more time.

Most of the returned participants (75%) thought the most important consideration should be the fact of shortages of personnel in their field, but only 54 percent thought it had been. Similarly, a majority (69%) said the academic record of the candidate should have been stressed; only 46 percent said that it had been stressed. Just half of the participants believed the economic importance to Indonesia should have been stressed; a minority (37%) said it had been stressed.

Some Kenteam comments sum it up: "Many were selected on the basis of individual ability and departmental needs; unfortunately others were selected on the basis of friendship and equally irrelevant criteria." "The drive toward excellence broke down much of the 'share-the-wealth' attitude."

Thoughtful views from IPB were expressed as follows:

The principal (in the first years) already recognized that students with the best chance for doctor's degrees should be extended. But it's hard to get the picture of their brilliancy because the reports of the U.S. institutions make nearly everybody eligible. If not grades then recommendations are always good. It's been difficult for the rector to decide even from the beginning on the question of extensions. There may be many deflections from the general policy. Also there is lack of manpower in certain fields and there is lagging interest in entering those fields. It is necessary sometimes to give opportunity to the less brilliant men. The "not so brilliant" students without Ph.D.'s may yet come home and carry out their work very nicely, because their organizing ability is good and this is a compensating factor. In this one can only talk about tendencies.

In extending programs the considerations were academic achievement, availability of that branch of science in Indonesia, the specific need of the department, the availability of staff in a department.

Majorities of Kenteam and IPB (70% in each case) noted that IPB, Kenteam, and the participants themselves experienced "rising expectations" throughout the course of the project, affecting—among other things—the length of time participants studied overseas, and Kenteam welcomed this development.

In the earlier years extensions were rarely granted because it was vital that Indonesians return to Indonesia as soon as possible to assume teaching responsibilities. As time went on, IPB wanted to upgrade their institution and therefore allowed students extension time in the states for advanced degrees.

More students became candidates for degrees. Considering the cost of getting them to the U.S. and back, the need for better trained professional staff, and the expansion of Indonesian staff so more members could be spared to pursue advanced training, this was a logical and economical request.

The Participants before Departure

The staff members who were sent overseas did not volunteer, but were nominated, examined for, and assigned to "study duty" (tugas beladjar). Objectives were designated for each person, and a faculty position was reserved for each to occupy upon return. The privilege imposed an obligation, so attitude and motivation were important, and these were special objects of inquiry in the survey of participants.

In recalling their feeling before going to the U.S., most of the participants said they had been eager for the experience or at least willing to go. Only a few claim to have been indifferent and none were opposed or reluctant. The most-professed motivation was patriotic—a desire to be of service in the development of Indonesia. Altruism ran high in the group, but self-interest was almost equally prominent. A large majority aspired to professional advancement. A majority also were curious: they wanted new experience, even adventure. Self-interest in economic betterment was expressed by a large minority (40%).

Preparation for the venture was a major concern of those who survived the selection process and were named to go abroad. They had to study the language and take examinations, decide what and where to study, make personal and family arrangements, get passports and visas, and be coached about life and work in the U.S. Kenteam and IPB members were heavily involved in these preparations, and Kenteam members especially made this a major interest and activity during their residence in Bogor. When the preparation of participants was abruptly halted by Indonesian decree in the spring of 1964, Kenteam members felt that their main opportunity to be of service had been shut off; they were visibly shocked and saddened.

Most of the participants (85%) recalled having received predeparture help from their Kenteam sponsors in various ways, more in filling out forms and preparing papers than in any other way. Most of them also recalled being helped by Kenteam in orientation to American ways; smaller proportions, in English language coaching, in deciding on what institution to attend and what study plans to make. However, about a third of the participants felt that they could have been helped even more by their Kenteam sponsors. IPB members advised as many participants on study plans as did Kenteam sponsors but were of less help in advising on what institution to attend.

Deciding on a Place and Plan of Study

Planning the study programs for participants was a matter of concern to everyone involved: the student himself, his colleagues at IPB, the members of the Kentucky team, the University of Kentucky, any other university which the participant might attend, and the Agency for International Development. There were so many interested parties that misunderstandings could easily occur. Almost half of Kenteam (47%), and more than half of IPB (67%) felt that there was a lack of understanding and communication between Americans and Indonesians in Bogor in working out study plans and future assignments for participants; even larger majorities of both groups saw lack of understanding and communication between departments in American universities and IPB departments.

In addition to the number of parties involved, there were other complicating factors. Long-range planning was new to IPB staff members and they often failed to appreciate the time required for preparations. In many cases there was no Indonesian colleague in his discipline whom a participant could consult, yet he could not be sure of either his needs or his wants. Often the specification of his future job was unclear.

Kenteam and IPB staff members advised participants on the choice of an institution and on the study plan before travel to the U.S., but final arrangements could be made only after arrival. These were confirmed, negotiated, or mediated by the contract office staff at the University of Kentucky. Problems arose if anticipated or desired study plans did not conform to available curriculum offerings in American institutions, or when admissions offices were perplexed by Indonesian academic records. In the choice of an institution at which to study, however, a majority of the participants (60%) recalled no problem. Two difficulties mentioned by a few participants were getting enough information about prospective institutions and finding institutions offering work in the fields desired. Incidental reference was made to a variety of other items, each a real "felt problem" to a few participants: avoiding a cold climate, finding another place after denial of admission by one graduate school, and reconciling the differing recommendation of IPB, Kenteam, and University of Kentucky advisers.

As indicated above, everybody concerned could have a hand in the choice of institutions at which participants should study but nearly half of the participants recalled making personal choices, and about the same proportion credited their Bogor Kenteam sponsors with decisions. Nearly a third recalled that UK personnel made decisions on institutions to attend. The IPB advisers were the least involved and were acknowledged as choice-makers by only about one-fifth of the participants. Choices of academic program and selection of institution to be attended often required compromise among the various relevant considerations, and there were some corresponding feelings of dissatisfaction. About one-third of the participants felt that the recommendations of their Kenteam sponsors in Bogor were not given enough attention in determining their academic programs in the U.S. Nearly equal proportions (19%, 21%, 22%) felt also that IPB's plan for the participant's study, the participant's own preference, and recommendations of their U.S. advisers were not considered adequately. Still fewer (13% in each case) indicated that their own previous academic record and U.S. graduate school requirements were given too little attention in the making of their academic arrangements.

Several participants reported changes in their objectives while in the United States. Only a few scaled their objectives down and many set their sights for longer stays in the U.S. One-third broadened their objective to include seeking an advanced degree, not originally intended. One-fourth attended another university than had been planned; they included 9 percent who attended the University of Kentucky and 15 percent who went elsewhere as a consequence of the changed objective. Some formed new economic objectives to save and invest or to buy "take-home" goods of exchange value in Indonesia.

The fact that program objectives changed for some participants after their arrival in the U.S. was sometimes a matter of concern to one or another of the parties of the affiliation. AID officials were inclined to feel that changes were proof that original planning had not been done carefully enough. Some IPB colleagues felt that changes would thwart specific plans made in advance for the work assignment after return from the U.S. But many of the changes were made on the advice of American supervisors of graduate study to promote the attainment of degrees. These supervisors resisted the recall of participants who were working successfully toward degrees but had not yet completed their work. The more or less fixed view of Kenteam members, generally supported by their IPB colleagues, was that each change in a participant's objective had to be judged individually. Generalized objection to change was unacceptable because the planning of advanced study must always involve flexibility and change based on periodic reassessment of the student and his experience. Kenteam members were often baffled by the inability of "bureaucrats and administrators" to understand that degrees are awarded for academic growth and accomplishment rather than for the passing of fixed time units. Nobody could predict how long a candidate must work for a particular degree and nobody could predict with certainty whether a given participant would "pass."

Orientation before Departure

Orientation is a subject of diverse views and feelings. It seems to be widely sought and approved by whoever is about to have a new cross-cultural experience, but it seems also to be widely criticized by those who are having the new experience after having been "oriented" for it. The difficulty seems to lie in errors of expectation. It is impossible to

118

replicate an experience in advance. Complete anticipation of all future needs is difficult, however urgently desired, and disillusionment after orientation is common. An interim conclusion may be stated to the effect that adjustments to a new experience are better and smoother if there has been some orientation; that orientation cannot anticipate all the new experiences to be faced; that the success of orientation is more a result of the effort made by its recipients than by instructors. It is evident also that interested and friendly "veterans" of experience, guides, and advisers can be helpful in preparation for cross-cultural ventures.

Among the most important predeparture coaches of participants, except of those who were the first to leave, were colleagues who had already returned from their own study tour in the U.S. They could speak with a type of authority that neither Kenteam nor ipb staff could match. The returnees were of the same status as the candidates except that they could speak from recent experience. Half or more of the participants acknowledged predeparture advice from returnees on how to travel to and in the U.S., where to live and what to eat, how to interpret Indonesian culture to Americans, how to fulfill the academic requirements of an American university, how to answer political questions about Indonesia, how to react to race discrimination in the U.S., how to save money, and how to plan a return trip. From one-fourth to one-half had been coached by returnees on how to associate with American friends and how to live on an American campus. One-third had been advised on how to buy and bring home a car.

Problems in the U.S.

After the stress of selection, preparation, and travel, there came eventually the moment of arrival in the U.S., the reception of the participant at the University of Kentucky, and the further planning of his studies in Lexington or elsewhere. Arrival was a critical event, fraught with both potential difficulty and promise. Queried about difficulties on arrival at the U.S. port of entry, however, half of the participants recalled none at all. Food and language problems were mentioned more times than difficulties in housing, facilities, and services. Only a few recalled feelings of fatigue or confusion and an occasional reference was made to difficulties involving climate, money, accidents, distractions, lack of friends, and time differences. Many recalled also having experienced language difficulty on arrival at the University of Kentucky, but there the other problems receded in importance and a new one appeared—that of academic procedure and planning. Some of those who did not remain at the University of Kentucky also encountered difficulties with academic plans and procedures on arrival at the university where they were to study.

None of these difficulties alone was of major importance as a cause of disturbance or frustration, but together they constitute a list of targets for planners of orientation programs.

Orientation after Arrival in the U.S.

All participants went through a prearranged orientation on arrival in Kentucky, but the nature and duration of the experience were not the same for all. For those arriving before 1962, orientation took no longer than two weeks, but a summer-long orientation was conducted at UK in 1962 and in each of the next two years. Orientation periods of less than one week, however, were not recalled as helpful in lessening the problems of language, academic adjustment, or social adjustment. Longer orientation periods were credited with helping with English, with adjustment to American university procedure and to life in the U.S. The "vote" was about two to one in favor of the eight weeks' orientation compared with periods of four weeks or less. That is to say, whereas nearly half of the participants (48%) reported having been helped by English study in the shorter periods, the participants were unanimous in feeling that an orientation period of eight weeks or longer had helped in English. A similar difference appears in the recollections about adjustment to university procedures and to life in the U.S. A number of participants offered miscellaneous comments on orientation periods. Some said "shorten or eliminate orientation," but others said "prolong it." Nobody called for less language instruction. Additional suggestions were for more visits, field trips, seminars, speech practice, and language instruction, provision for more participation by the students themselves in the planning of orientation, deeper study of American culture, and more attention to university procedure and practice.

When initial problems had been overcome, others came to attention: illness, lack of news from home, academic problems, language, homesickness, international matters, lack of contact with IPB. Only two of these problems were cited by a majority of the participants, the first being homesickness, acknowledged by just half of them. Next came lack of contact with IPB (44%); then there were academic problems (18%); language difficulty (16%); international affairs (13%); lack of news from home (13%); and sickness (12%).

Participants reported various efforts to prevent or quell homesickness: they wrote many letters home, attended football and basketball games, watched television, kept in the company of fellow Indonesians, studied long hours, took part in sports, attended movies, and organized parties. Frequency of movie attendance was at least once a month for nearly all (90%) and once a week or oftener for some (21%). Participants interpreted Indonesian culture in many ways while in the U.S.; most of them (75%) demonstrated features of dress, music, and dance, and a majority (60%) also reported having made speeches at meetings of organizations. Ten percent joined honorary societies at the American universities where they studied. Although policy on both the Indonesian and American sides of the affiliation prevented their acceptance of other fellowships or assistantships, at least 3 percent were offered assistantships to encourage their further study and to solicit their participation in research at the offering institutions.

Allowances and Their Use

The expenses of participants during their period of authorized residence in the U.S. were met by monthly allowances and by reimbursement for officially approved disbursements. Initially the allowances were fixed and uniform; later they were fixed in relation to cost-of-living indexes of the communities in which the participants lived. The modal percentage of the allowances that was required to pay for food lay between 50 and 59; two-thirds of the participants (66%) estimated that their food costs consumed between 40 and 70 percent of their total allowance. No estimates were requested for other budget items except savings and only 12 percent did not answer, or reported no savings. Half of the participants (55%) reported they were able to save 50 percent or more of their allowance. More than a third bought automobiles; about the same proportion kept savings in money form; several bought radios, records, and record players; others bought clothing beyond immediate needs. Others (10%) bought typewriters, cameras, pens, watches, slides for projectors; another 9 percent bought sewing machines or kitchenware. Six percent bought books.

A major economy practiced by participants while in the U.S. was in the preparation of their own food (reported by 41%), very often cooperatively with other Indonesians (reported by 54%). While this reduced their costs of subsistence, economy was no doubt only one of the reasons. Even more important was their firm commitment to a rice preference, reinforced in many, perhaps in most cases by the traditional and widely prevalent pork-eating tabu.

Social Contacts

Americans in Kentucky frequently commented on the fact that the Indonesian participants were inclined to live with each other and to prepare much of their own food. In addition to the satisfaction of eating rice, their staple food at home, this enabled them to converse in the Indonesian language and to enjoy the fellowship of their countrymen. However, this also gave them some reputation of social aloofness within the academic community. Only 9 percent reported *not* having roomed with other Indonesians. Experience in living with one fellow Indonesian was reported by 15 percent, but groups of three or more were more common. In fact, 64 percent had dwelt with two, three, or four room or house mates. In response to another question, 58 percent of the returnees confessed to a wish that they had learned more than they did about American life. The same proportion said they would have learned more if they had followed some other arrangement with respect to dwelling.

Various kinds of housing were reported, with the rented apartment the most common (reported by 81%). College dormitories came next (41%). Twenty-one percent had rented rooms and 16 percent had lived in the homes of American families. All had lived with others; only 18 percent had lived alone at any time in the U.S. When asked if they would live in the same way if in the U.S. again, 68 percent said yes. Some of those who replied no were apartment dwellers who said they would prefer to live with an American family. The preference for apartment living was based on the desire for privacy, freedom to follow their own eating and religious customs, economy, the possibility of reciprocating the hospitality of American friends, and practice in being self-sufficient.

The average participant had visited as a guest in thirteen or more homes. Only 6 percent had been a guest in fewer than four American homes, and 22 percent had been in more than twenty such homes. Forty percent reported eating in an American home once a month, 35 percent twice a month, and 10 percent three times a month; 7 percent said they ate with American friends in the home of the latter at least once a week. Yet it is remarkable that as many as 3 percent reported having never eaten as a guest in an American home.

Perhaps more important than association with American friends were contacts maintained by correspondence to and from family members, friends, and colleagues "back home," halfway around the world, in Indonesia. However, 10 percent did not report receiving, and 13 percent did not report sending, letters from or to family members in Indonesia. One-fourth of the participants received and sent letters home once or twice a month. Two-thirds exchanged letters three or more times a month. Nearly half (41%) reported not receiving letters from IPB colleagues, and even more (54%) reported receiving no letters from Kenteam sponsors at Bogor. Only one participant reported receiving letters oftener than twice a month from IPB colleagues.

Absence from Wives

Many participants were absent from their wives during the period of overseas study, and this was a matter of concern to both Indonesian and American parties to the program.

A majority of the participants (75%) were married, but of these 25 percent were newlyweds, married within six months of their departure date, as often in conformance to the norms of their parents and the community as in response to their own wishes. Forty-five percent had been married a year or more; 19 percent five years or more. More than half of these participants were separated from their wives by travel and residence overseas for more than a year, and 13 percent were separated for more than two years.

Eight percent of the husbands had their wives with them in the U.S. for periods of six to fourteen months. In only one case did the wife travel to the U.S. with her husband. In the other cases, wives followed when their husbands had saved enough money, or had otherwise arranged to finance the travel and costs of living.

When asked if it would have been desirable to have their wives with them in the U.S., 80 percent said yes; only 10 percent said no. The enforced absence of husbands and wives was a negative feature of the participant program.

The wives with overseas husbands had to make special living arrangements. Twenty percent maintained their own households in a rented or owned house. A larger percentage (42%) lived with relatives in Indonesia (not necessarily in Bogor) and the remaining 38 percent lived with others in Bogor or elsewhere. Half of the wives left behind were engaged in some kind of work for pay—17 percent taught, 15 percent did secretarial work, 7 percent did sewing, and 10 percent had a variety of other jobs. Their social activities included participation in faculty wives' organizations (55%), student wives' organizations, churches (15%), and various other groupings.

Participants as Graduate Students

Indonesian participants generally were in good repute in the graduate schools they attended, as indicated by the response to a questionnaire addressed to the professors who had been their advisers. The most interesting information from this source concerned the adequacy of the participants' preparation at the time their studies in the U.S. began. A majority—but not all—were reported to have been well enough prepared for reading, writing, and conversation in English. Only 16 percent were judged to have been initially deficient in reading ability, but about one-fourth (27%) were thought to be deficient in writing skill and about the same proportion (25%) in conversational ability. The deficiency percentages in these language skills were largest for the social science students, whose field required more proficiency in language, and least for those in the physical sciences. One-fourth of the participants (23%) were said to have had deficiencies in basic science preparation and a slightly larger percentage (28%) were remembered as being deficient in their understanding of scientific method. These lacks were reported for a smaller percentage of social science and a larger percentage of physical science students. It may be concluded that most of the participants were adequately prepared in English and in orientation to science but that better preparation would certainly have expedited their progress through graduate school.

The tables turned, however, when advisers were asked about the industry, diligence, and motivation of their Indonesian students. Nearly two-thirds were rated superior and only 8 percent deficient in these qualities. None of the social science students were rated as "lacking in this respect."

Rigidity of Graduate School Requirements

One complaint sometimes heard at IPB was that graduate school requirements in American universities rigidified the study programs of participants to their disadvantage. The American advisers felt, however, that graduate school requirements had imposed hardship on not more than 5 percent of the participants in the recognition of credit for undergraduate work. But 60 percent had been required to take additional prerequisite courses. Nearly half of Kenteam (47%) and about the same percentage of IPB staff (44%) agreed with the statement that graduate study for IPB staff members in American universities forced them to follow rigid degree programs that did not prepare them specifically for their work later on.

Indonesian Data at American Universities

It seemed to some at IPB that an American university could not understand the special needs of Indonesians for graduate study. The United States was so far away. One problem, it was thought, would be lack of Indonesian data at American universities. Less than a third of Kenteam (31%), but more of IPB (43%) thought this might be a problem. IPB members registered a little more concern on this point than did Kenteam; it was they who experienced or feared the deficiency—in different forms. One form was the lack of familiarity of American graduate school advisers with Indonesian academic credentials; another was the lack of Indonesian materials for study at American universities. But a common view was expressed by a member of the IPB staff, who wrote, "Why should one pursue graduate study in the U.S. limited to Indonesian conditions only? What is important is not the subject matter but the research method. Even if one wants to limit oneself to Indonesian conditions, in some fields one could obtain more material about Indonesia in the states than in Indonesia (at IPB)." A Kenteam member put it this way: "In most cases the students were not working on Indonesian problems and therefore did not need Indonesian data. For general cultural reasons and for the development of the American universities, however, it would have been important to have collections of Indonesian library materials. This is a particularly bad deficiency at the University of Kentucky."

Performance and Standing

In general, advisers claimed that they expected neither more nor less of Indonesians than of other graduate students but held to the same standards of work. For only 10 percent of the participants did the advisers confess leniency, requiring less than of other graduate students. Stating this in a slightly different way, 17 percent of the advisers confessed that they graded Indonesian graduate students more leniently than others.

Indication of successful study comes from an analysis of the first-semester standing of the participants, although this was difficult to ascertain because of the variety of time schedules and evaluation scales used in the several American universities concerned. Translation was effected to a common scale from 4 (A) at the high end to 1 (D) at the low end. In their second and later terms, 68 percent of the participants were performing at A and B levels.

Comparison of Participants at UK and Other Institutions

In general, the estimates and judgments of participants as reported by their advisers are similarly distributed among those who attended the University of Kentucky and those who attended other institutions. There were a few minor differences. At UK there was a slightly larger percentage of participants in the social sciences and a smaller percentage in applied agricultural fields. More than half of the UK participants, less than half of the others, remained in the U.S. for longer than twelve months. The percentage getting master's degrees was larger at UK, the percentage getting Ph.D. degrees was greater elsewhere. There was no difference between UK and other universities in what advisers "expected" in the work of participants. But 20 percent of the participants at UK and 16 percent at other universities were reportedly graded more leniently than other graduate students. The proportion with a standing of 3 (B) was greater at UK; the proportions with grades lower and higher than B were a little larger at other universities. Just half of the participants got higher standing in their second term than in the first, and the figure was slightly higher at other universities than at UK.

There were no significant differences between UK participants and those at other institutions in the impact of graduate school requirements. However, at UK there were slightly larger proportions of nondegree participants whose work was of graduate level in quality but who did not have a degree objective or a long enough period of U.S. residence.

Opinions, in Retrospect

Except for political opposition during peaks of anti-Western emphasis, there was widespread IPB approval of the participant program. Skepticism was occasionally expressed, however, and the following discussion presents Kenteam and IPB opinion with sample comments on the criticisms made.

A majority (62%) of the participants who had returned to Bogor before Kenteam's departure had been promoted one or two ranks above their previous level. Several others had returned recently and might expect promotion as soon as the internal developmental processes at IPB had organized the necessary departmental and college arrangements.

There was a tendency among both Kenteam and IPB members to evaluate participant training against somewhat unrealistic expectations that returnees would make immediate, full, and fruitful application of their new knowledge upon return to duty in Bogor. While overseas, absent members had been missed by family and faculty; upon return they were met affectionately by family but somewhat querulously by faculty. It was not everywhere understood that the benefit of foreign study, like that of any advanced study, is revealed over a period of time, not all at once. It was not everywhere understood that the greatest benefit is in the continuance of intellectual growth which eventually, but perhaps not immediately, finds practical application in problem-solving and crops out in teaching, in research, and in service to community. But the time of evaluation began immediately by colleagues who had waited at home and by all who had an interest in the outcome of the training program.

One IPB staff member complained that study abroad made the participants less, not more, able to meet expectations of the Indonesian people in regard to national progress and development. On this question, Kenteam and IPB members' judgments differed slightly, but among the Kenteam and IPB opinions combined, there were two confirmations of "more able" for each one of "less able"; in other words, a two-to-one endorsement of the participant program. A representative Kenteam view was, "I think the one most important thing they see in the U.S. is that the individual can have an effect in bringing about change (contrary to insja Allah [God-willing], mungkin [maybe], barankali [perhaps]). And they learn some of the means to do this."

A majority of Kenteam (67%) and an even larger majority of IPB (who were really in a better position to know) agreed that participants had taken initiative and demonstrated self-confidence by introducing new courses, changing course content, and using new teaching methods on their return from overseas study. Half or more, by their own report, had undertaken to use various new teaching methods: making use of libraries, requiring reading, allowing interruption of lectures with questions, organizing field trips, assigning term papers, using laboratory teaching, giving frequent quizzes, adopting informality in lecturing. Not so many (but between 30% and 50%) of the returnees were assigning homework, assigning grades along the range of percentiles, using objective tests, and determining grades from "the curve."

Incidental other features of American college and university organization and activity had been learned by the participants, especially aspects of departmental organization and administration (by 40%) and college organization and administration (by 29%). University-level organization and student organization had been especially noted by 18 percent.

The Kenteam majority who thought returned participants were starting new research activity (59%) was smaller than the IPB majority (90%). That returnees were also increasing their activity in extension and community service was agreed by less than a majority of Kenteam (39%) and a somewhat larger proportion (46%) of IPB colleagues. On all three of the foregoing questions, however, many Kenteam members abstained, no doubt mainly those who served in the earlier years and did not know returned participants. These responses indicate that early evaluation was running in the returnee's favor, and that those Kenteam and IPB colleagues who proffered their opinions agreed that changes were being introduced in teaching, research was being started, and extension and community service activities were being undertaken by the recipients of overseas education. A Kenteam member commented that

On first arrival until the last year or so participants have returned with enthusiasm and a spark in their eye apparently eager to try out some of the things they had learned. They have of course gone through initial periods of disillusion and frustration. The constant succession of crises, both in the society outside and internally, has slowed down the process of their assimilation. A chronic shortage of money together with some lack of attention to planning is involved. In the last year and a half, especially the last year, participants have returned to a situation of increased political and economic uncertainty, and the application of their learning in the States probably awaits a more fortunate situation. I expect their demonstration of initiative and self-confidence to be delayed for two or three years, but I expect it will come out in the open later. So far as research is concerned they particularly suffer from lack of resources, facilities, and money.

It's not so much the book-learning as observing first-hand that there are ways of working and organization which are effective, and that individual initiative does play an important role in making progress. Seeing is believing and ideally they have some understanding of the context in which things are written.

One critic of overseas study, who had not been happy in his own American experience, expressed the opinion that an IPB staff member could learn as much from American scholars at a distance—from journal articles and books—as from going there to study. Kenteam members were in complete disagreement, as were a majority of the IPB staff (57%).

A view once expressed by one of the senior professors at IPB was that in their postgraduate study in the agricultural sciences, regular IPB staff members should learn research methods and procedures by working only on Indonesian agricultural problems, studying and writing only in the Indonesian language. Kenteam and IPB disagreed sharply on the matter, the prevailing view of the former being opposition and that of the latter approval, in full or in part. The one-fourth of Americans who approved felt it was true "for the time being, because the need of Indonesian agriculture is so great and money so scarce for research; for political reasons also." "They should work on applied problems; the research method can be learned in many places, including Indonesia. But the scientific language of Indonesia should be English. It is the most important one in the world now; to use Indonesian would be a step backward."

A resolution of the apparent differences of view between Kenteam and IPB on the point can be seen in a research and graduate study procedure now gaining favor among cross-cultural educators. In general this involves baccalaureate training in the native country, introductory postgraduate study (course work and a qualifying exam, if for the Ph.D.) in another country, and return to the native country for data collection, followed by thesis writing and examinations in either the native country or the foreign country, depending on where the graduate advisers and examiners find it possible to be on duty at the time. It is this kind of graduate training that organized foundations and cross-cultural agencies for education and development may be implored to support on an expanding scale.

One highly placed member of the IPB staff complained that returned participants showed they had learned research technology at a craftsman's level but failed to make application to problems related specifically to Indonesia's plans for development. Kenteam members and IPB staff were asked whether they agreed. Kenteam either did not know (42%)—again because many of them had not been in Bogor when participants returned—or agreed only in part (36%); a majority of IPB (67%) agreed in full or in part. One Kenteam member said, "Failure of the returned participant in doing research is not necessarily caused by failure to make applications to Indonesian problems. He might be overburdened by administrative and teaching jobs. I still have to find a research worker who still remains productive in his field though he has a teaching load of twenty hours and a consulting job to do."

The Results of the Participant Program

Discussion continues among international technical assistance personnel over whether it is wise to send representatives from developing countries in the tropics to the United States for advanced technical and scientific study, especially in agriculture. It is asked whether the content of their learning is relevant to current need in their own countries; whether the necessary prolonged absence diverts their career interest.

The participant program at Bogor, however, was a massive input to the development of IPB. Other parts of the institution-building effort would have been less effective, perhaps almost futile, without it in the face of the particular threats that had to be met. The curriculum could diversify and broaden only as participants returned to staff the necessary teaching posts. Departments and colleges could form only as chairmen and deans could be named. The participant program provided the solution of the moment to one of IPB's problems of staffing, and none of its other developmental needs were more urgent than this. The presence and organization of its participant-trained staff were IPB's major defense against institutional erosion when "the going got tough" in 1965 and during the next few years.

Had "the situation" in Bogor been more favorable, returning participants could have contributed much more—and more quickly—to the institution's development and to helping it find a destiny in the nation's growth. Unfortunately they could not make ends meet by working for IPB alone, and everyone had to "moonlight." For some, their major effort went into seeking supplementary income, and their service to IPB was hardly more than nominal. But they were there; IPB had a staff "on board," with reserve potential whenever its full energies could be applied.

Aside from these emergency features of the particular situation at Bogor, several general factors account for delayed satisfaction of the usual expectations of foreign study. The first is the perplexity a returnee feels when he seeks to connect his newly learned theory and method to the "real world" in his own country. The next is the felt "strain toward consistency" which permeates the group of colleagues he reenters after an absence which may have fostered some alienation. Some of them were not privileged to study abroad and may covet what little control they exercise over returning upstarts! There is a strain to keep the work and life of the returnee from getting too "far out" and to make them consistent with the routine of the day. These and similar deterrents delay but do not stop the forward movement of the returnee from advanced study abroad if he was well chosen, suitably guided, and effective in his self-effort, as were most of those from Bogor. There is always a post-return pause which many observers have neither the insight to identify nor the patience to endure. Their criticism of the returnee is like complaining about a newborn baby's loss of birth weight before starting again to gain! The pause in professional effort and accomplishment on the part of a participant may be brief—a few weeks—or it may last even a few years, but the chances are good that eventually he will make the vital connection between his own education and his country's problems, that he will get past the stage of trying to transfer the elements in his foreign experience and will reach the stage of trying to adapt them and of creating new intellectual compounds.

The weakness in the participant program at Bogor lay in its termination rather than in the features noted above. IPB was frozen at a half-way stop. Only half of the returned participants had even a master's degree and the number who received the Ph.D. could be counted on one's fingers. This was hardly more than a start up the long climb to competence and even less progress toward excellence. Restrained from further progress by the stoppage of 1966, IPB could hardly maintain the ground it had gained without further inputs from outside. Deteriorations in equipment and facilities, the slow-down in library growth, and the sapping of strength by secondary jobs made continuation without deterioration almost impossible.

A check-up made by Kenteam's last chief of party during a brief visit to Indonesia in May and June 1968 indicated that nineteen participants had not returned or were still in foreign study, and one was deceased. Of the 184 returnees, 96 percent were professionally active in Indonesia. A majority, 70 percent, were in service at IPB, but 26 percent were engaged at other posts. By consideration of IPB self-interest, this reveals some brain-drain, in that only three out of every four professionally active returnees are serving the institution for which the contract intended to train them. But the fourth returnee, who has moved into some other position, is not a complete loss to IPB; he is at another college of agriculture, or working in one of the ministries; a few even are in some kind of agribusiness. Thus many of the losses to the IPB faculty are gains in linkages with other agencies and organizations also at work in Indonesian development; the investment in their advanced education has not come to rest in IPB but it is still potentially productive in the furtherance of progress in Indonesian agriculture. The loss of returned participants from the IPB faculties is evidence of "normal" professional mobility in a nation's allocation of manpower to jobs.

6. Kenteam, the Agency for International Development, and the University of Kentucky

Kenteam and AID

The contract between the University of Kentucky and the United States Agency for International Development (formerly the International Cooperation Administration)[1] established the broad purpose and the terms of work of the University of Kentucky's representatives—Kenteam—in Bogor. As in all such contracts during that period it was the responsibility of a chief of party to direct the team's work. The responsibility of a division chief in AID/Djakarta was to administer the agreed support of Kenteam and to assure that Kenteam's work was within contractual authorization. For a portion of the period Kenteam's contract was with the agricultural division of AID, but during most of the years it was with the education division, which maintained operating relationships with the Indonesian Ministry of Education. The Kenteam project was an integral part of the American technical assistance program in Indonesia. AID/Djakarta was the executor of the program established by Congress, authorized from year to year, and financed annually by an appropriations bill, with allocations through the U.S. Bureau of the Budget, the president, and the AID director in Washington, and coming eventually to the overseas mission.

Kenteam and the AID division were partners in an enterprise of joint interest and responsibility. There were, of course, ebbs and flows of understanding and misunderstanding because Kenteam members were professors, whereas AID members were officials in a bureaucracy, and all of them were a long way from home. Between professors and bureaucrats there are usually small—and sometimes large—credibility gaps. The tensions that play between them are normally stimulating and supportive; issues and moments of conflict, however, are not uncommon because each sees any problem from the vantage point of his own work and his own habit. Kenteam felt sometimes neglected and sometimes over-supervised; AID felt sometimes avoided and sometimes pestered!

Bogor and Djakarta were then from one to two or more hours apart, depending on road conditions and traffic. So Kenteam members and families, living nearer than university teams in other cities, were frequently in the national capital and usually found occasion to drop in at AID and the embassy. Kenteam's chief of party and administrative officer attended the regular meetings of the AID directors and division heads in Djakarta. Djakarta Americans came to Bogor much less often, although a few commuting AID families lived in Bogor during some of the period. Personal and interfamily relationships between Kenteam and AID were cordial, but, with a few exceptions, more casual than intimate. Each group had its own neighborhood and spent most of its social time within its own circle. In Djakarta there were more Americans and they lived in a metropolis; in Bogor, fewer Americans in a smaller city associated readily with Indonesian colleagues and had mixed circles of friends. In 1960 there were thirty-four American families in Bogor; this was probably the maximum number during the period of the affiliation. At the end there were only about ten.

AID's interest in Kenteam was always keen and generally affirmative. Changes in "the situation" in Indonesia, changes in AID program emphasis and personnel, and changes in Kenteam personnel required each party to undergo frequent reorientation and to foster continuing renewal of acquaintance with problems, programs, and rules. During the nine years there were two ambassadors, three AID directors, three chiefs of AID agriculture, three chiefs of AID education, and three Kenteam chiefs of party; none who were present at Kenteam's arrival were present at departure. Only among Indonesians, mainly in Bogor and Djakarta, were there any who could have

[1] See John Gardner, *AID and the Universities* (New York: Education and World Affairs, 1964).

taken part in the whole experience, and even they could not have known the American sequence. This is one part of the story of any AID mission and any university contract group. The Kenteam-AID relationship is significant as a general rather than a unique matter, even though special features were present, and even though reorganizations modified the character of AID after 1966.

Bureaucratic elements and the detail of logistics may be treated "once over, lightly." Kenteam sought more funds than could be provided; AID sought economies. Kenteam members sought and eventually were granted rights of commissary purchase and APO mailing. They also sought eligibility for health and medical facilities that were not at all times accessible. They sought all the forms of compensation and allowances provided by the contract agreements and sometimes resisted bureaucratically administered restrictions. They sought all possible securities for life and work in Indonesia where "hardship post" supplements were provided. AID representatives sought always to assure that transactions were within the rules. Kenteam guarded jealously its responsibility for educational programming; AID sometimes sought to utilize the expertise of Kenteam personnel for other AID purposes. One difference was in orientation to the use of time. Professors think of teaching hours, credit hours, semesters, study time, research time, paper-grading time, and academic years, and they have an academic concept of leave. Bureaucrats think of the work day, the work week, month, and year, office hours, overtime, and annual leave. To apply the bureaucratic definition to the academic habit was always a problem. The two groups shared to some extent an operating jargonese, but AID retained superior skill in communication by PIO/T, PIO/C, and PIO/P (project implementation orders for technicians, commodities, and participants). AID officials were at home with the ProAg (program agreement) E-1 (the plan of work), PPA/s (project proposal and approval summary), and TDD (terminal delivery date). To them, a professor was a technician. Kenteam especially resented alleged instances of "changing the rules during the game"—making new interpretations from Lexington, Washington, or Djakarta, of rules—

such as of those applying to the allowable class of international travel, homeward shipment of cars bought overseas, "third-country" travel privileges, and freight weights. To the professors, an AID-auditor was an off-side judge of educational programs and a foreign service inspector was among the least welcome of many visitors. AID employees were jealous of Kenteam perquisites, including U.S. income tax exemption. Kenteam was jealous of AID perquisites, including reimbursement of travel for "R and R" (rest and recreation) from service in a "hardship post."

Kenteam's visitors over the years included the ambassadors and others from the embassy, the AID directors and division chiefs, AID representatives from Washington, survey teams, a cabinet officer, at least three senators, and annually representatives from the University of Kentucky. Generally, the hospitality of Kenteam, the friendliness of Kenteam's Indonesian colleagues, and the program briefings for visitors were rewarded by approval and occasionally even commendation. At times, Kenteam felt its work to be among the least-favored of AID activities in Indonesia; but as the years passed it achieved more American recognition and acknowledgment, and the two Kentucky teams (Bogor and Bandung) outstayed the other AID groups and were withdrawn from Indonesia several months after other projects ended in the turbulences of 1965 and 1966.

AID's chief quarrels with Kenteam were over alleged delays in planning; oversight and error in observing administrative regulations; too little concern for the American taxpayer's money; incomplete communication and infrequent reporting or reporting that lacked dynamic thrust; inadequate justifications for requested expenditures; laxity in requiring that commodities purchased with AID funds be marked with AID emblems; allowing equipment purchased with AID funds to fall into disrepair and nonuse; slow ordering and delivery of commodity purchases; tardiness in shifting American teachers out of basic sciences and into applied agriculture; and seeming aloofness with respect to other AID projects. There was one period of extreme tension in which an AID division chief threatened severance of relations, charging University of

Kentucky neglect of the project by providing only part-time or incompetent campus coordinating personnel. At another time AID was considering enlargement of Kenteam and extension of its coverage to several other institutions.

Kenteam's chief quarrels with AID were in cases of what it considered intervention in educational programming decisions within agreed budget limitations. After dropping a team position in farm management to add one in agricultural marketing, in a normal progression of programming, Kenteam was chided in a letter from an acting director of AID for depriving the team of an economist. A marine biologist was recruited by the university for Kenteam and AID announced that no such position was authorized. A plan of work was prepared by AID without reference to certain written suggestions from Kenteam. An appreciative AID director frightened Kenteam by stating his wish to call upon its members as though they were his own staff in an AID division of agriculture. An AID director asserted his view that Kenteam was too large a group to keep at one institution and should be scattered, one or two professors to each of several other faculties. A judgment was made that Kenteam was too sophisticated and was building an ivory tower at Bogor, pouring graduates into bottomless pits and neglecting needy outpost-institutions by passing over demands for middle-level personnel. A division chief visited Bogor and went individually to each team member without clearance from the chief of party. An AID criticism, without relation to program needs, objected to degree-training for certain participants who were following planned study programs in American universities. These kinds of proposals were viewed by Kenteam to be out of context, requiring responses that could be made only within the framework of joint IPB-Kenteam program study.

But these were all problems within an arena of common obligation and commitment. The contract and its amendments defined the arena and the rules of action. Operating relationships between AID and Kenteam provided a supportive structure of administration and educational programming under an umbrella of American foreign aid, and there was more solidarity than alienation between the Americans in Bogor and those in Djakarta. The AID chiefs of education were pivotal figures in this pattern, mediating for Kenteam the pressures from congressional reviews and the top-side bureaucratic structure, seeking to wrest from Washington more field control of university contracts, and linking Kenteam to the American system of agencies and provisions for foreign aid. To Kenteam, then, the chief of the AID education division was sometimes a nuisance, sometimes a benefactor, but always a friend.

Kenteam and the University of Kentucky

There are everywhere a number of common attitudes among underdeveloped peoples. Many problems in the Southern Appalachians are similar to those in Asia. Hence, the capacity of the University of Kentucky for work in developing countries had been tested by long experience in the development of its own region. The posture of the university had been shaped especially by a real concern for the mountain people which applied with equal relevance to problems in other lands.

Toward the end of World War II a president of the university, Herman L. Donovan, had said, "Our campus is the State of Kentucky," and in this he reaffirmed a long-standing statewide obligation. Only a few years later, his successor, Frank G. Dickey, enlarged the concept to include service in other countries, extending statewide to worldwide. The university's first venture abroad had been in Guatemala, where technical assistance was undertaken in agricultural education, with support from the U.S. International Cooperation Administration. The second was half way around the world in Indonesia. Kentucky had no concentration of Southeast Asian competence, and one opinion among many academic thinkers is that nobody should serve in a region without first achieving area expertise. However, to require pre-established area specialization would limit the overseas work of American universities to only a few institutions or a few departments and faculty members therein, and to only a few parts of the world. There is some concentration of competence in science and technology at every university and it should be a part of every modern university's character to explore new fields of educational and developmental service. The University of Kentucky, though not an institution of world renown, was at least in the middle range of sizes and qualities among universities in the United States. In Southeast Asian cultural knowledge, it was underdeveloped, in tropical agriculture it was inexperienced, but in general agricultural skill and comprehension, in research, teaching, and extension experience in its own developing hinterland, the University of Kentucky was not without superior potential for service in new and unfamiliar areas. Located within an ethnocentric American region, with a tradition of working mainly in its own community but taking note of the post-World War II period of global American involvement, the University of Kentucky would have been in default of its educational purpose had it not been willing to try foreign service, partly to help with development in remote places and partly to advance its own development and improve its own capacities. As for the requirement of prior area expertise, it may be considered useful, but not essential if workers in a new area apply the basic approaches of men of knowledge, seeking to learn new things in the effort to apply what they already know.

Kenteam at Bogor was an overseas educational unit, with the size of a department and the diversity in function of a college, responsible administratively to a special office of the university. Initially this special office was the Kentucky Research Foundation (KRF), the contractor with AID in behalf of its parent institution. In the early years, KRF was well known in Bogor for its relationship with KCT (the Kentucky Contract Team—Kenteam) at the two Bogor faculties (later IPB) of the UI (University of Indonesia).

But to many Indonesians, KRF was lost in the thickets of corporate organization. Many of them, as well as many functionaries in AID and many faculty members in Lexington, questioned the nature of KRF's involvement. Appropriate as a contracting unit for the university, with certain flexibility of administration in the face of state legalities, KRF was thought to be an unsuitable unit for the operation of educational programs. The director of KRF served part-time as campus coordinator and had a staff to serve the contract. The staff recruited professors for overseas, programmed the study of participants, processed orders for equipment, generally monitored the contract, kept books, and communicated with the team in the field. It may be that in putting forward the symbol of KRF the university did not establish a proper image with any of the other parties to the contract until the later years of the affiliation.

After certain complaints were received from AID/Djakarta, after a heightening of tension between Kenteam and KRF, and after the disaffection of university departments, especially in the College of Agriculture, steps were taken to pull the management of the contract more closely into the university's educational program. An experimental step was taken when the contract was five years old—in 1962—and returned team members were added to the KRF staff to assist in campus coordination. The dean of the graduate school became also the director of KRF in 1962. In the same year an Office of Overseas Programs (UKOOP) was established, responsible to the executive vice-president of the university. This placed the campus management clearly within the university rather than in a semiexternal unit. This arrangement was further confirmed by assigning the campus coordinatorship to a newly organized Center for Developmental Change in 1966. But through the years, there was a failure to engage the full and dedicated interest of university departments, especially those in the College of Agriculture, the university's most competent unit for staffing the program in Bogor. It was a weakness of Kenteam, Bogor, that it didn't have enough members from university departments and that only the last of three party chiefs was a University of Kentucky professor and former department head. The university learned well some lessons from these ventures, and later, with the agreement of AID, adjusted its next overseas commitment (to work with an Agricultural Center in northeast Thailand) to safeguard against repetition of some of the mistakes made in organizing the work of Kenteam at Bogor.

Once the contract had been formalized, stating the purpose

of the program and fixing the terms under which the university would act as contractor, a major responsibility of the coordinating office was to staff Kenteam, prepare members for duty, and send them to their posts. The chief of party (at first called the team leader) and the campus coordinator were the line of connection between the university and each team member while abroad, whether or not the member was previously in a university department. It was as though each member was on leave from his department. This, however, was a mistaken conception. It should have been—and at the end it was recognized—that Kenteam was not on leave but on duty for the university overseas instead of at home, and that rights of full membership in the university continued in force and rights in the future continued to accrue.

Tensions between Kenteam and KRF developed and subsided in alternation. Most issues were resolved but some were more or less continuous. The files reveal numerous complaints from Kenteam directed to the coordinator's office—complaints about KRF's part-time attention to Kenteam (the whole program); major decision-making by minor personnel; the indifference of university departments to Kenteam members; alleged failure to back up Kenteam in differences with AID in Djakarta or Washington; overpromising and underwarning new members; alleged deficiencies in orientation; delays in purchasing; "hang-ups" over placement of participants in American universities and unilateral changes in study programs of Indonesian participants in the U.S.; ambiguous explanations of rules; infrequent or incomplete communication of information; assignment of uninvolved university administrators to make inspection visits; violations of lines of communication; noncooperation in recall of a team member unsuited to his work; lack of clarity in fixing responsibility for programming and budgeting; failure to allow Kenteam its share in responsibility for programming and budgeting; failure to develop an on-campus educational program; failure to give attention to returning members in order to make use of their expertise; occasional use by subordinates in the coordinating office of whip-snapping language and grumbling to admonish and scold; lack of understanding of Kenteam's medical and health needs, vacation and travel procedure, and various lacks in the provision of amenities; inadequate attention to suggested extensions of tour for Kenteam members and extensions of study leave for participants; KRF's reluctance to let Kenteam correspond with American universities to plan participant study; and Lexington's difficulty in understanding sensitive aspects of "the situation" in Indonesia.

There were some things that Kenteam wanted but could not have or did not get for some reason or other. Kenteam wanted rupiah funds for local research; it wanted American graduate students to do research in Indonesia; it wanted full exchange at teacher levels, with Indonesians going to the U.S. not just as graduate students but some as visiting scholars or guest lecturers there. Kenteam wanted third-country travel for themselves and their colleagues; they wanted to make purchases from other than American vendors; they wanted participants to stay longer in the U.S. to get degrees; they wanted salary increases and promotions to come along as readily while they were abroad as they would at home; they wanted language lessons; they wanted to hire a school teacher; they wanted the chief of party to visit Lexington in reciprocation of inspections from the university; and they wanted an occasional department head to come as an inspector.

Similarly, there were issues generated in the reactions of the campus coordinating office to Kenteam. These issues, too, were mainly resolved as time passed and wisdom grew. Lexington's chief unrest with Kenteam was over delayed or incomplete communication. It took at least twenty days (before APO, which reduced the time to five to seven days) for an exchange of letters. Telephone calls from Bogor were only occasionally satisfactory and it took as much planning as for a Fourth of July picnic to complete a call. Telegrams were prompt but often garbled and more difficult to translate than Indonesian news. The Lexington staff wanted better job specifications for use in recruiting team members. They complained of careless purchase orders, incomplete information about participants, neglect of follow-through on getting

various Indonesian approvals, stubborn and tenacious adherence to demands for privileges of travel, leave, various
allowances, disinterest of team members who were not from
the University of Kentucky in promoting the university.
Lexington felt that Kenteam was sometimes lax in reporting
and failed to keep the university up to date on changes.
Lexington responded acidly to the acrimony in certain letters
from irate chiefs of party, and the files document a few
first-class quarrels. These, by the way, occasion a remark
about what may be called deflected aggression. The inside
members of a group, such as a family or a department or a
team, must sometimes react in their own explosion chamber
in order to keep aggressions from getting out of hand in
other directions, as from Kenteam to AID or IPB, or from the
university to AID.

Orientation

Orientation of team members for their work in Indonesia
was a responsibility of the university's coordinating office,
but orientation, especially if conducted casually, is a game
that nobody wins. The orienter concludes that it is impossible
to prepare his clients properly without more time and resources; the oriented concludes that he is really not well
enough prepared and should have learned more; members of
the host community decide that the orientation was too
shallow and ineffective. "Why didn't they tell me about
this before I came here?" was a common reaction of newly
arrived Kenteam families. But anyone's first response to a
new cross-cultural situation is as much a function of his
personal readiness as of the culture he is entering. Readiness
is attitudinal as much as informational and the requisite
attitudes are deeply lodged in personality, beyond suscep-
tibility to fundamental change by contrived acts of orientation.

Eventually Kenteam members came to see that really no
one could possibly have anticipated and explained all their
future needs and problems. Realizing something is more than
hearing about it, and life anywhere is partly a process of
revising one's working illusions. Any perplexed newcomer

finds it easy to blame everybody who didn't "tell" him in advance.

Kenteam members with previous experience abroad were more confident in their initial response to Indonesia, Bogor, and IPB than others who were having their first foreign adventure. Half of the Kenteam members were experienced in life and work outside of the United States, and half were not, the work at Bogor being their first abroad. About 20 percent had European experience and 26 percent had previous experience in Asia. Only a few had been in the Middle East (8%), Africa (5%), or South America (8%). Previous work had been in military service, teaching or lecturing, survey or research work, consulting, technical assistance, or study. Military experience, consulting, and technical assistance were the most frequent. Several Kenteam members had third-country travel adventures, both professional and touristic, en route to or from Indonesia or on trips during their tours—in Australia and New Zealand, Japan, the Philippines, Malaya, Thailand, Taiwan, India, places in Europe. Some of those visits were coordinated with study trips of IPB staff members who had completed study overseas and were making visits on their return trip to Indonesia.

It was a fairly uniform view among all Kenteam members, however, that a more intensive orientation to Indonesia should have been arranged for them before departure from the U.S. Orientation in Lexington for most of Kenteam consisted of a few conversations; for some, a few lectures, a little mimeographed material, some suggested reading, and no language instruction whatsoever. Especially in the earlier years, before any veterans of service in Bogor had returned to Lexington and before any Kenteam tradition existed, there were elements of overpersuasion and overpromising in the recruitment process by zealous agents of KRF and the office of campus coordination. There were cases of subsequent resentment at alleged unfilled promises, some with justification, others probably without foundation. A few members had thought they might become chiefs of party but other things always happened. AID in those days had little interest in reimbursing universities for orientation; they had little awareness of the need or confidence in its efficacy and they wanted to get on with the job. When Kenteam first went to Bogor, *The Ugly American*, a semi-fictional criticism of Americans abroad, had not yet been published and the need for screening out poor prospects, for cultural instruction and language training were not as widely acknowledged then as later.

In a survey made in December 1965 Kenteam members and their families mentioned several procedures which they felt could have been followed by the University of Kentucky, by AID, by their Kenteam predecessors and colleagues, or by themselves and others, for more effective orientation. Their check-list for the University of Kentucky included the use of senior and full-time personnel for coordination at the university; more reliance on the chief of party for help in selecting overseas staff; better job descriptions; more time for preparation (orientation, language study); one month of "duty status" for intensive language study before departure; clarification of regulations and one authoritative informant on rules; more information on details of the contract; salary increases to reward language proficiency; more factual information on living and housing overseas; written statements on the aims and history of the project; the introduction of newly recruited members to on-campus departments; the use of returned Kenteam members to help orient outgoing members; less "sales pitch" in recruitment; and more realistic assessment of field conditions.

They suggested that AID could have contributed to the better preparation of team members by arranging for consultations by new staff members with other universities and private agencies before going over; by providing intensive language study; by providing orientation to customs, religion, and life in Indonesia; by making available a small library of relevant documents; by encouraging professional side trips en route to and from Indonesia; by providing information on AID activities in that country; by briefing sessions with persons familiar with local conditions; and by fuller explanation of regulations.

Kenteam predecessors and colleagues, it was thought,

could have contributed to the better preparation of new members by more thorough briefing; help with the language; better job descriptions; introductions to personnel in Indonesia; preparation of housing before arrival; advance ordering of supplies and equipment; direct communication from the chief of party before arrival; better orientation in Bogor: visits to other faculties and institutes, and more information about departments, courses, and other characteristics of the host institution.

Kenteam members thought of a long list of things that might have been done by the host institution to prepare them better for their work: better orientation to the IPB system; explanations of goals and objectives; introductions to colleagues; suggested reading lists; a better description of the assignment; greater frankness from the beginning; regular inclusion in departmental meetings; more conferences; and more liaison with government ministries.

Kenteam members also saw in retrospect many things they could have done for themselves in the interest of better preparation. They thought they should have read more books on Indonesia, tried harder to learn the language and something of the culture of Indonesia, talked more with people who had been there. They could have had more communication with returned members and participants in the U.S. They thought they should have insisted on better job descriptions and better understanding of the contract and work situation. They could have made better preparation for taking books and equipment. They could have undertaken more professional training (restudying introductory crop, soils, and plant taxonomy materials, for example). They felt they should have done more "calling" in the early stages of their tour, making more use of the entree which any newcomer can exploit. They could have taken more initiative from the beginning in Bogor, based on a better understanding of Indonesian customs and attitudes. They should have been more open-minded and compassionate . . . and . . . !

Kenteam was equally divided on the answer to the question: "After arrival in Bogor were you told what was expected of you?" Half said yes, the other half said no. Clearly, more could have been done by way of clarifying the expectations of Kenteam members before and at the time of their arrival, and in the early days of their work in Bogor. In this connection, a small majority of Kenteam members felt that they understood the USAID-UK contract, which established the terms for their service in Bogor, but 44 percent felt they did not understand. There were mixed feelings of uneasiness; on this point Kenteam members were less than satisfied in their understanding of the contract. Better communication and firmer continuity of policy would have been helpful.

Concluding Note

The experience of other universities during the Kenteam period has been the subject of several studies. As a result of these, tendencies have been put in motion that will change the relationships of a group in the field to the foreign assistance agency and to the home university in future projects. The Gardner Report initiated a review that has continued; many issues have been analyzed and recommendations made in the CIC-AID research report, the Pittsburgh group's work on institution-building, and the study by special committees of the Agency for International Development and the National Association of Universities and Land Grant Colleges.[2] As this book approaches publication, there are portents of many innovations to come.

As for Kenteam and the University of Kentucky whose interaction has been explored above, it has to be observed that one was part of the other and that the lists of issues no doubt exaggerate an impression of relationship problems. Throughout the whole period the two were mutually supportive; the files and reports confirm their basic harmony and agreement in purpose and procedure. Issues and snags did exist and they were the problems that had to be worked on; harmony takes care of itself.

As for Kenteam and AID, the end of every quarrel was achieved by simple recourse to full communication. When an

---

[2] Bibliographic references to these reports appear in the *Report of the Joint Committee on AID-University Relationships* (Washington, D.C.: Agency for International Development, 1970).

issue was fully exposed and explained, it could be and was understood and acceptable decisions eventually were made. There were sometimes disappointments on one or both sides; the professors and bureaucrats never completely exchanged roles, but a working equilibrium was maintained. There may never be a time when all credibility gaps between educators and bureaucrats can be fully closed. Their interaction abroad follows patterns that are well known at home, but probably with more concentration and intensity of stimulus and response at a remote post.

7. Images of Kenteam: Self and Ascribed

Observers have wondered whether professors from temperate zones in the West can be helpful as guest professors in the tropics of the East. Can they bridge the wide gaps in ecology and culture with the hard material of scientific universals and technology and with the softer material of their own goodwill and dedication? For Kenteam there are data by which to seek some answers. Accumulated files, now in the archives, yield abundant information from correspondence, minutes, record forms, memoranda, reports, and diarylike notes. Additionally, there are data from a small-scale survey conducted by interview and questionnaire which provide a measure of Kenteam's estimate of some of its own capabilities, and parallel measures of the same Kenteam qualities viewed by Indonesian colleagues. The survey questions were presented to prospective respondents in Bogor in December 1965, a time of maximum sensitivity and tension in general relationships between Indonesia and the U.S. They could be used only selectively and with consideration for the fears and uncertainties of the moment. This was a post-coup period when many Indonesians suspected even their friends, when political forces were seething, largely beyond the view of Americans, who were isolated within rather than threatened by the swirl. The intensity of vital and immediate concerns, the urgency of preserving personal security, the effort of the university leadership to keep IPB at work and on target could allow only low priority to a questionnaire. The IPB respondents, however, either by interview or with written answers, included all present and former deans, rectors, associate rectors, and presidium members then living in Indonesia. These thirty-two Indonesian respondents were thoughtful persons and their responses identify prominent features of the affiliation. Responses of the guest professors (Kenteam), of course, were another matter: of a total of forty-seven, thirty-eight (81%) returned the schedule.

The reflection of Kenteam's self-image was sought in each member's estimate of the capabilities and characteristics of his colleagues. No member was acquainted with all others who served between 1957 and 1966, so each could be asked for his judgment of only the colleagues he knew. In a sense, then, each member had his own team and there were forty-seven Kenteams! The Kenteam self-image is a composite of the colleague-estimate of all members who "voted," that is, filled out the questionnaires.

The image of Kenteam held by colleagues in IPB—the ascribed image as distinct from the self-image—was similarly determined. Each of the thirty-two-member panel expressed his characterization of the Kenteam members he knew, so the ascribed image is a composite of thirty-two Kenteams. The percentage was computed of the self-group (Kenteam) or others-group (IPB) who judged that most Kenteam members "had" a designated quality or capability.[1] These percentages represent Kenteam's two images and their identity or difference. In the differences between the self-appraisal and the evaluation-by-others may lie some lessons about doubt and confidence in the conduct of cross-cultural technical assistance.

A Mixture of Qualities

The general, over-all summary evaluation was easier to make than separate judgments trait-by-trait. Ninety-two percent of the Kenteam members thought their team had a "good mix" of qualities.[2] A typical view was, "Yes, it was a good mix; about as good as one would encounter in a department of similar size in an American university. Usually some

[1] Features of the Kenteam images are discussed here in the order of their presentation in Appendix B, Table I; that is, from the highest to the lowest percentage of Kenteam members who agreed that Kenteam "had" the characteristic.

[2] Cf. Harland Cleveland and others, *The Overseas Americans* (New York: McGraw-Hill Book Co., 1960), pp. 124 ff. The concept of "good mix" is discussed here in the meaning used by Cleveland and his associates.

members of a department are superior, others are just 'riders.'" ipb was a little less generous but a majority, 67 percent, credited Kenteam with a "good mix." The response to this question in general terms was less discriminating and hence somewhat more superficial than the responses given one by one for specific items.

Competence

Hardly anyone, American or Indonesian, disagreed with the statement that Kenteam members were sufficiently qualified in technical skill.[3] They were all, of course, accredited by their employer, the University of Kentucky, and were accepted for work in Indonesia after an ipb review of their qualifications. All had been educated in their specialties in American universities, and 70 percent of them were doctors of philosophy. The self-image of Kenteam and the ipb ascribed image differed only in degree of certainty; more of Kenteam agreed in full, more of ipb agreed in part.

Kenteam thus considered itself generally competent and Indonesian colleagues agreed. There was recognition in both groups, however, that this competence could have become more specific to the ipb situation if some Kenteam members had known more about the tropics to begin with, had been more effective in communication, had been more versatile in working out Indonesianized applications, and had not confined their work so narrowly within their specialties.

The Department-Member Role

Were the Americans of Kenteam careful to recognize their roles as department members, acknowledging the authority of Indonesian department heads and deans? A booby trap lurks in the use of the role concept in this application—the danger that roles will be thought of as parts in a play that the actor may learn and for which the lines and the props and the postures are based on pretend rather than on "real life." Some extroverts, marginal men, and "fast movers" may become so facile in shifting among the roles of a big reper-

[3] Ibid., p. 128.

toire that their self is only a kind of kaleidoscope of facades. They are the "too smooth" type! Role playing may be done consciously in a play or unconsciously in the depth of daily living, and it makes a lot of difference which.

Eighty percent of Kenteam and over 63 percent of ipb expressed the view that the department-member role was correctly played, so this was the dominant view. Comments of some ipb staff reflected the circumstance that departmental and college structures were in process of formation, and that clear definitions of role for staff members were not always evident. The ambivalence revealed in answers to this question was probably inevitable for two reasons. One was the incompleteness of ipb's organization at each stage; this meant that ipb and Kenteam staff alike were uncertain of their own stance, of what programs to pursue and with what persistence. The other reason was the generally uncertain expectation that an American and an Indonesian had of each other. Each desired that an effective relationship would develop between them and each was restrained from entering such a relationship fully by his uncertainty about acceptable degrees of assertion and inquiry, and appropriate forms of response.

The difference between the Kenteam self-image and the ipb ascribed image may point to a cultural difference in expectations. Indonesian practice normally involved a little more subordination to department heads than American customs require.

Organizational Skill

Most Kenteam members credited a majority of their colleagues with organizational ability, but most members of the ipb panel thought only a minority of their Kenteam colleagues had this kind of skill.[4]

There was clearly a variety of meanings in the minds of Kenteam and ipb staff members: organization of time schedule, lectures, labs, equipment orders, home life, departments, courses, relationships with colleagues. It was pointed out that "organizational ability stateside is not necessarily organizational ability in Indonesia. Our food technologist, our inland

---

[4] Ibid.

fisheries specialist, our virologists, our plant physiologist, our chemist and physicist organized their laboratories. The difficulty was in organizing IPB staff and student personnel for full understanding and use of these laboratories. Kenteam was all right on organization of technical work; it was a little bit short on organizing for effective communication and acceptance by Indonesia colleagues." Another reason for the difference between the self-image and the ascribed image no doubt lies in the fact that Kenteam members thought of themselves as having been charged with an organizational responsibility, whereas IPB personnel thought of organization as a function of their own prerogative and for their own initiative and decision.

Commitment to American Educational Methods

Are American educators working in host institutions abroad too strongly committed to promoting American concepts of education, including college organization, curriculum content, and teaching methods? The question was asked about Kenteam and its work in Bogor. A large majority of Kenteam members (72%) and a small majority of IPB members (56%) concluded that most Kenteam members were *not* overcommitted to American ways. This is tantamount to asserting that they were flexible, capable of adaptation and change, and would not impose American procedures, even though taking all their major cues from their experiences as Americans. Kenteam comments explaining their views in this matter were expressed rather fully in remarks more extended than responses to some of the other questions. This invites two explanations: one, that they were defensive; the other, that education was their business and that they had studied carefully the question of how much American educational procedure to promote. A representative expression of Kenteam's view of itself on this point was:

Initially it was probably the intention of each Kenteam member to start right off with the first step: one, two, three to replicate American methods in American institutions. Fortunately, this did not last very long in most of the Kenteam members. They were flexible enough to see the need for developing Indonesian patterns, and fortunately the official policy of the team from the first chief of party on through was to help the Indonesians develop what they wanted by way of an institution rather than to impose the American system. In actual practice, however, an American teacher tends to go on teaching with his use of a syllabus, his lectures, his illustrations, his library assignments, his quizzing procedure, according to his habit as an established teacher in the U.S. Probably it is not so much a matter of promoting as it is of continuing to live according to habit.

A sobering Indonesian response was:

Nearly all stressed too much the system with which they were familiar. This is natural because it is difficult to switch to another system with which one is not familiar. It was not so much that they were "pushing," but only that they were used to their own system and naturally considered it better. In the early stages of the affiliation this was felt more than now because IPB participants after returning began to use more of the American system which they had learned, so prevalence has gradually grown.

Aggressiveness

Debates raged—and they will probably continue in history—between proponents of the hard line and those of the soft line in American-Indonesian diplomacy. Ambassador Howard Jones was criticized for being "too soft" in his representations to Sukarno; his defenders asserted, however, that he was "soft" in a tough way! There were moments also in Kenteam when patience ebbed and the temptation was to be "hard" or "tough" in the approval or disapproval of expenditures, of participant nominations, and so forth. By Kenteam criteria, however, the "soft line" philosophy prevailed throughout the affiliation. But IPB and Kenteam sensings of softness were not identically calibrated; there were times when Kenteam thought itself "soft"—but IPB's reading was "tough." Some IPB comments argued that Kenteam dominated commodity purchasing, Kenteam composition, and even the selection of IPB staff for training abroad.

Do Americans among colleagues in another society refrain too much from pushing for changes they think are needed, or do they "push too hard"? One of the image-probing questions about Kenteam was on this point and there were many (more than one-fourth) of each group who abstained, unsure enough of their view to record it. This suggests some confusion about how much aggressiveness would be appropriate in the context of the Kenteam-IPB relationship. Yet a majority of Kenteam (69%) and almost a majority of IPB (47%) were of the opinion that most Kenteam members would not have accomplished more by more insistent methods of promoting changes. The Kenteam self-view was a little more favorable to aggressive effort than the IPB view. Clearly there was diversity among team members in the force and nature of pushing. One discerning Kenteam comment was, "In the Indonesian setting, I think the surest way to stop progress is to push—not only is the American resented, but communication comes to a standstill. Traditional ways of working have to be considered, and although the Indonesian way is terribly frustrating for an American, I think the team members did well in taking things slowly and working for understanding rather than instant progress." An Indonesian viewpoint refined the meaning of "pushing": "It isn't the amount of pushing that is important here but the manner of pushing. Nobody likes the idea of being pushed. . . . One good policy was the concept that the team did not have a policy of its own, but offered ideas and services in a certain direction."

Political Skill

An acknowledged requisite for success by Americans in work overseas is the sense of politics, explained as political sensitivity and tact.[5] Whether or not Kenteam members had "political savvy" elicited differences of opinion within Kenteam itself and between Kenteam and IPB. This was another topic on which many of the Indonesians (30%) abstained, being either reluctant to express themselves or uncertain.

A majority of Kenteam (61%) but only a minority of IPB (21%) were of the view that most Kenteam members had political skill. Certainly the individual members of Kenteam varied greatly in their political manner. Kenteam and IPB

[5] Ibid.

were thinking of two very different political arenas when responding to the question on "sense of politics." One was the internal, academic-affairs arena of the university itself; the other was the large and vigorously active arena of the ongoing Indonesian revolution. IPB was less positive than Kenteam that Kenteam had the proper "sense of politics." As with several of the other attributes it was agreed that there were members with no, some, much, or very much political skill. The IPB panel's judgment was "skewed to the left" and Kenteam's self-estimates were "skewed to the right." Americans were thinking mainly that the appropriate skill was to be nonpolitical, both within the faculty and in the surrounding political milieu. Indonesians were thinking that Americans should be neither politically active nor politically inept, and some of them felt they were the latter.

Understanding Goals

It is an orthodox requirement that change agents understand the aims, purposes, and goals of their host community or institution. "The most important single step in establishing a climate for effective communication is to try to understand and appreciate the community's goals for itself, and to make them in some real degree [one's] own."[6]

How well does a group of university professors achieve and exemplify this requirement? Do technicians, moving cross-culturally and with a two-year tour of duty, succeed in understanding the goals of their colleagues?

Kenteam gave itself a better score on this trait than that ascribed by IPB. A majority of the former, though not a large majority (58%), and only a minority of the latter (30%) concluded that most Kenteam members acquired the requisite understanding and appreciation, in terms of the above quotation. Expressing this in a further free translation: the American view was that most Kenteam members understood Indonesian goals; the IPB view was that most Kenteam members did *not* understand.

No explicit statement of IPB's goals was ever given to Ken-

[6] Ward H. Goodenough, *Cooperation in Change* (New York: Russell Sage Foundation, 1963), p. 388.

158

team. In general, the goals were left to each team member to discover, assume, or infer. It was reported that the Dutch goal had been only to serve their plantations and to replicate for this purpose their agricultural university at Wageningen. American officials had described *their* goals—not necessarily those of IPB—in documenting the official project statements, and they had American land-grant models in mind. Seemingly, the Indonesians wanted mainly to make the institution their own as quickly as possible and to make the institute a servant of national independence and development. At Bogor, the lack of clearly formulated and announced specific goals was nobody's fault; it was only a mark of the distance yet to go in developmental action.

Various goals were made evident in various ways and at various times—goals *of* IPB, which IPB established for itself, or goals made *for* IPB externally by various persons concerned. For some the goal was to make IPB a "mother institution" in Indonesia; or to make it a center of excellence in Southeast Asia; or to completely Indonesianize IPB; to make IPB an effective force in increasing food production; to train some announced number of insinjurs per year to work for the ministries and estates; to modernize the curriculum; to achieve autonomous university status. There was an unsorted mixture of remote and intermediate goals. As one Kenteam member said, "goal formulation was taking place all the time. It took too much time perhaps for the Americans to discover what the goals were. All Americans accepted the goals without difficulty."

But there also was some doubt that Americans *could* understand the aims of an Indonesian institution. One IPB staff member found the question "very difficult to answer since I sometimes wonder if *I* understand IPB's goals." Another thought Kenteam understood "as well as any outsider can understand the goals of a struggling university where the staff works at huge odds and not for money but for a way of life."

There was indifference, even irritation on the part of some Kenteam and IPB members alike at the implication that there should be any responsibility other than a "concern primarily with one's own discipline." In the words of one Kenteam member,

There was a tendency for each team member in the beginning to assume that the goals of IPB were like those of any agricultural college such as those for whom he had worked in the U.S. In nearly every case this view was later modified. Most team members came to understand IPB's obligation to take part in nation-building and to participate in the Indonesian Revolution. A smaller number came to have some understanding of the nature of the Indonesian Revolution. A still smaller number achieved a deeper level of understanding in appreciating IPB goals, accepting them as valid, and assimilating them into the team member's own attitudes.

Indonesian doubt of American understanding was expressed in a few special comments, such as:

They all understood and accepted IPB's goals but it did not mean that they understood Indonesian ways of achieving goals.

I believe that almost all Kenteam members understood or came to understand and work by IPB's goals but it takes some doing (and time) to adapt oneself to a system or a line of thinking that is basically different from that which one is used to; adaptation is more difficult when it comes to the details of executing this system. How much time and effort was given to preparing Kenteam members before they came to Indonesia? IPB policy fits into Republic of Indonesia policy; they are the same. Had Kenteam members been prepared to understand this?

In reviewing these statements on Kenteam's understanding and acceptance of IPB's goals, no major failures or breakdowns can be identified in the relationship described, but one may indeed derive a lesson. The initiation of planned, directed, and intentional development must not overlook the search for and distillation of goals. They must be the goals of the host, but the guest can help to explicate them, and in any event the guest's helpfulness will be limited if the host's goals are neither understandable nor acceptable to him. Formulation, clarification, and acceptance of goals are first orders of business.

Directness of Approach

Would Americans in Asia—or elsewhere—meet with better response if they weren't so "direct" in making suggestions? This question may overlap in meaning, or in a sense be another formulation of the question above about "pushing" for change. One-fourth of Kenteam members and half of IPB members failed to respond to this question. Perhaps the Indonesians found the question perplexing because it offered no cue as to the definitions of direct or indirect. For Americans, the question is important because they are sometimes baffled by the slow and circuitous routes of approach to decision-making which they encounter in developing societies. They have forgotten the historical past in which indirectness in communication was a feature also of American rural life. Comments by Kenteam members reflect their deep interest in the question, and their effort to understand.

Relatively few Americans without previous overseas experience had the patience and wisdom to "make haste slowly."

To an American a suggestion *is* just a suggestion; to an Indonesian it may be a command or a criticism.

It seems to be an art to make suggestions indirectly but effectively, or very directly without offending.

Custom and etiquette here carry the implication that to challenge somebody's idea directly is insulting to that person. Our custom is to try to treat ideas objectively as something to be examined from all points of view much as one would examine a piece of merchandise before buying it. This leads to hurt feelings at times.

It is difficult to tell when Indonesians are offended. Undoubtedly, several good ideas were passed over because of the way they were presented by the Kenteam member.

I found in most cases a suggestion went farther if it was given in the form of several alternatives so that the colleague could then make a selection.

Only a few comments were written in explanation of IPB views. They suggest that "indirectness" per se may not always be the problem, that the quality of the directness may also be relevant.

In my opinion constructive suggestions, even "direct" ones, have always been appreciated, especially if they are made in a way that doesn't show any shadow of superiority complex. It's not the "directness" that may cause a less favorable response, but rather "the way in which" and the "timing."

I think that the Kenteam members should make direct suggestions. We can accept these from a foreigner, but not from an Indonesian colleague.

The Americans are too discreet in making suggestions. It is indeed true that in society in Indonesia, Americans have to comply with the current official policy-line—often contrary to their own personal opinions. But whether this is always the right thing to do is very questionable.

Kenteam acquired some but never enough skill in ways to avoid or relieve embarrassment on the part of Indonesian colleagues. Direct criticism could only be painful; direct rejection and denial were always hurtful. To reverse oneself was to lose face and a change about could be accomplished only by rearranging the elements in the problem situation and dealing with the rearrangement through an intermediary. These rules of face-saving were as simple and easy to understand as the rules by which Kenteam members had always behaved; it was just that, to some extent, they were different, and there was neither time nor opportunity to acquire familiarity enough to be competent in their exercise.

Teaching Methods

This question is closely related to the earlier topic of commitment to American ways, and both are given more extended attention in another chapter of this book, so not much more than the "vote" need be presented here. But one of the most notable changes at IPB during the early years of Kenteam's presence was the introduction of what Indonesians refer to as "guided study," which involved different methods of teaching and evaluating students, in contrast to former "free study" practices. The teaching component in Kenteam was a dominant feature of the Kenteam image, and IPB gave stronger endorsement to Kenteam's teaching methods than did Kenteam itself. In answer to the question of whether teaching methods introduced during the Kenteam period were an improvement over former practices a majority of Kenteam (58%) but a much larger majority of IPB (80%) agreed that the new methods were better.

The main vote was clearly a view that there had been improvement, but there was also a mixture of uncertainty, partial agreement, and even a little disagreement which no doubt reflects the uneven adoption of changes. The direction was toward required and regular attendance, continuous residence, the use of textbooks and reference materials, the use of quizzes and written examinations, the preparation of written reports, relatively more reliance on laboratory learning and relatively less on lectures. When Kenteam left Bogor, however, none of these changes was complete, and in some parts of IPB there was some return to former practice—more in response to shortages of books and manpower than from rejection of the procedures. The Kenteam view was that "the change has not yet gone far enough."

Travel

A fringe benefit of working in Indonesia was the chance to travel there. All Kenteam members traveled internally, some a lot and some a little. Estimates were that they had spent an average of forty-five days in internal travel. During the entire period, 195 trip-units[7] were reported.

Only a few of Kenteam had limited their travel to West Java, the home island of Bogor and Djakarta. Aside from points in West Java the most visited areas were Central Java (perhaps because the easiest to reach) and Bali, perhaps because reputedly the most exotic of all spots. Nearly everyone had gone by car across Java, or by plane from Djakarta. Seventy percent had traveled in East Java. Next in order of

[7] A trip-unit was a single trip, or a group of trips described in one report paragraph because they were all to the same place and for the same purpose.

reported visits by Kenteam members were Sumatra, Sulawesi, Madura, and Kalimantan.

How much should an American travel in a "new country" before getting down to business in technical assistance? Would Kenteam members have been more effective if they had traveled more? Their qualification-by-travel was a feature of their image; bare majorities of both Kenteam (56%) and IPB (51%) expressed their judgment that Kenteam had traveled enough in Indonesia. But nearly as many either abstained or registered doubt.

A great variety of objectives and experiences is seen in a catalog of the reports of Kenteam members. Included are tours for observation and familiarization; study trips with classes; visits to students on field assignments collecting research material; observation of field surveys in progress; visits to enterprises, agencies, experiment stations, research institutes, colleges, universities, and agricultural schools; visits to rubber, tea, coffee, and chincona plantations, oil refineries, fertilizer plants, textile mills, craft centers, transmigration projects, sugar factories, mechanization projects, rice projects, tidewater areas, pine forests, logging locations, and corn experiments; trips to collect soil fungi, observe grasses, study livestock feed, note the effects of pests and diseases, observe veterinary field services, collect plankton and marine animals, see fisheries installations, visit field extension activities, and observe cropping systems and shifting cultivation; trips to give lectures, attend academy of science meetings, survey other institutions, and carry out special missions; and, last but not least, just plain tourism, sight-seeing, and recreation.

Neutrality

Where Americans work, there are contending viewpoints and groups; factions are in conflict with each other and visitors can readily choose sides and join. No doubt a newcomer may conclude that one or another viewpoint is "right" or better than opposing views, and it becomes difficult to remain aloof in the battles over adoption of practices and preferences for persons. An acknowledged principle of technical assistance—or of counselling, advising, doing social work, negotiating—is that a change-agent must avoid becoming embroiled in the internal conflicts and factional disputes in the client community, and those who know colleges and universities know how much they are sometimes ravaged by struggles within. Certainly it behooves a visiting professor to tread cautiously in the no-man's-land between the factions.

Did Kenteam manage this problem well? A majority of Kenteam members (56%), but only a minority of IPB (26%), expressed agreement that most Kenteam members kept clear enough of factions. Nearly a third (30%) skipped this question, being uncertain or unwilling to respond. The comments of IPB staff members indicate that some of them thought the questions referred to factions within Kenteam rather than within IPB. One IPB member observed, "This is probably one drawback of a team composed of recruits of several universities. To eliminate this kind of rivalry one ought to try to let the future team members get themselves acquainted first in the U.S."

It seems clear that Kenteam members responded only in terms of IPB factions. The following is a representative explanation of the American viewpoint:

There was a tendency for most Kenteam members to identify with certain Indonesian staff members with whom they worked most closely. Hence, there was a tendency to become involved in the same disputes or activities that the counterparts found important. To the extent that issues concerned academic matters such as curriculum, course content, time schedules, etc., it was useful for Kenteam members to have a voice in discussion. But when it came to who was going to be rector, or who would be members of Dewan Curator, or whether IPB should continue as a part of the University of Indonesia or be made separate and autonomous, or who was going to be dean of animal husbandry, the situation became different, and generally speaking, Kenteam members were aloof from the conflicts that were engendered.

Methods of Evaluating Students

The ways in which members of different societies evaluate each other are as different as their general norms and specific customs of measurement. Students in the pre-Kenteam days in Bogor were graded by annual examinations given orally, and in case of failure were reexamined as often as desired. As one IPB member stated, "Formerly we had the Dutch free system of study with only 10 to 20 percent of the students passing a course each year. Under the new system, we can pass 70 to 80 percent."

Kenteam and "guided" education introduced changes which are discussed more completely in another chapter of this book. Available to note, however, are the Kenteam self-view and the IPB view of whether the newer methods of student evaluation were an improvement. A majority of Kenteam (54%) and an even larger majority of the IPB panel of respondents (67%) considered the changes in student-evaluation procedure to be better than former methods.

Sense of Mission

A small majority (53%) of Kenteam members and less than a majority (40%) of IPB respondents identified a "sense of mission" as a characteristic of most Kenteam members. The definition in hand was taken from Cleveland: "A belief in mission is something more than a willingness to work in foreign countries for a long period of time. Among both career and non-career personnel some are enthusiastic about their jobs, some have a sense of purpose and achievement, and some regularly carry their work far beyond the call of duty."[8]

IPB respondents made fewer explanatory comments to support their votes, expressing viewpoints similar in meaning to those of Kenteam, but weaker in tone. There was a tinge of feeling that Indonesians expected more, or something different from "what they got." There was also some criticism. Of one Kenteam member it was said by IPB, "He is more like a Dutch professor than any other Kenteam member we've ever had." Of another it was thought that he gave more thought to Christian missionary work than to technical assistance. Another was thought to be totally ineffective and concerned only with personal comfort and gain; another was criticized

[8] Cleveland, *Overseas Americans,* p. 131.

for interventionism; another was thought to be maneuvering for a longer stay in Bogor rather than for the development of IPB.

The Kenteam images are generally clear with respect to sense of mission. The impression is one of a Kenteam which did its work and which had more members with than without this attribute. This impression is further confirmed by their special comments on motivation. Factors that had attracted members to Kenteam, in order of mention, were the chance to travel and see a new part of the world, the chance to join an educational effort to make improvements in an under-developed part of the world, the chance to get a change in job and family living conditions, the chance of a bigger salary and the opportunity to save more by living less expensively, and the chance to get ahead, be promoted, and advance professionally. Travel opportunity and the chance to join an educational effort tied for first mention. However, change in job and bigger salary were each given first importance by two or three Kenteam members.

Empathy

Empathy seems often to be understood as a synonym of sympathy. But empathy is an unemotional ability to understand another; sympathy is an emotional response. Empathy is skill in imagining oneself in the other's place and in imagining the other in one's own place. It is an automatic or habitual response that keeps one aware of others, a device for improving one's accuracy of understanding. If one speaks of a friendly or unfriendly sentiment or of pity or envy, one speaks of sympathy, not empathy. In fact, the development of sympathy with its emotionality may at times get in the way of the clarity and objectivity in understanding produced by empathy. Empathy—not sympathy—is the requisite tool of a change-agent, although he will not, as a personality in his own right, be devoid of sympathies.

Kenteam's capacity for empathy appeared in about the same manner as did its sense of mission, and the meanings of these concepts must have mingled in the minds of respondents. This definition also was borrowed from Cleveland: "The skill to understand the inner character and meaning of other ways of life, plus the restraint not to judge them as bad because they are different from one's own ways."[9]

It was a fifty-fifty opinion with Kenteam members, and nearly the same with IPB (47%), that most Kenteam members were equipped with the social-psychological radar known as empathy.

Inhibitory Effect

An aspect of Kenteam's images was its "valence" as a facilitating or inhibiting influence on IPB activity. A problem of team members was to discover relationships, joint activities, and kinds of participation in which their influence would be effective and positive. Did they find it useful and effective to take part in the rounds of meetings at IPB? Or did their presence in departmental, faculty, and other meetings reduce the effectiveness of those meetings? If Indonesian colleagues refrained from full expression of views in the presence of Americans, if American expression was awaited and depended upon, if language barriers resulted in misunderstanding, or if undue time was required for interpretation or repetition in a second language, it might be expected that American presence would indeed slow things down. There was no regular provision of interpreters, and Indonesian staff viewed themselves and were viewed by each other as professional colleagues whose attention needed to be given to the substance of their work. It was a common experience for Kenteam Americans to sit through meetings with only a partial and summary knowledge of what transpired, unless the meeting was conducted in English or unless a friend took the trouble to explain in English.

However, fewer than a majority (45%) of Kenteam members and an even smaller percentage of IPB staff (38%) felt that Kenteam's presence in meetings was inhibitory. A representative viewpoint from both groups might be called the "it depends" attitude.

From Kenteam:

[9] Ibid., p. 136.

168

Yes, language was one problem; respect for seniors was another; reluctance by Indonesians to discuss their problems in the presence of Americans was another problem.

When such meetings were conducted in English, I'm sure some of the Indonesians were inhibited. I know of one who refused to participate.

I think that this depends a great deal on the team member. I'm sure that at least one or two had such rapport with their departments that they did not inhibit participation by Indonesian colleagues. I did not have this rapport, for they always expected me to say something, and unfortunately, I usually did.

From IPB:

There are two things that should be taken into account: a) there are the so-called internal Indonesian meetings just like Kenteam members have among themselves; b) if the invited Kenteam member does not speak Indonesian then the whole group is more or less compelled to use English which for several people may be both embarrassing and less fruitful.

Yes, but this was not because of language. Presence itself inhibits—the Indonesians would refuse to resolve the small differences between them in the presence of foreigners.

Underutilization

It has been widely observed that experts abroad seem to be underutilized, in the sense that either their abilities are not fully recognized by their clients, or they are not called upon or not permitted to operate in many situations. Kenteam members did, at times, feel that their energies and abilities were not fully exploited by IPB. The survey indicated that less than half of Kenteam (45%) and even fewer IPB colleagues (30%) felt that the abilities of most Kenteam members were fully utilized. But the matter was clearly not perceived in the same terms by the Americans and the Indonesians; there was more agreement on the fact of under-utilization than on the reasons for it.

Some Kenteam members felt it was their own lack of assertiveness: "I don't think it's necessary to wait for an IPB request before at least planting the seeds of one's own ideas." Another view blamed IPB's failure to recognize the broadness of the American's training (even an agronomist has had some work in animal husbandry, and in the latter field a poultry specialist has general knowledge of all the major livestock classes). It was felt that IPB was too specialty-conscious, both in accepting the work of guest professors and in planning advanced training for its own personnel. As participants returned with degrees, however, this problem became less acute because the new Indonesian leadership came to understand more about the actual breadth of U.S. education by their own immersion in it. It was felt that "our people all had ability in many areas other than their specialties. We could have done much more in sanitation, in nutrition, and in helping the Indonesians live easier lives if our technical know-how had been exploited."

The American view was that IPB and its staff did not have the absorptive capacity for greater service from Kenteam. "Practically all of us could have done more in or around our subjects or fields; however, the facilities of IPB and the time available to Indonesians would in large part have prevented any such extension." "I know of no one who made all the contributions he or she was able to make on a broad scale."

Indonesian viewpoints stress other factors. There was the view that Kenteam was *not* underutilized "by Indonesian standards." "Serving more broadly could have meant dominating." On the other hand there was agreement that "every specialist has at least a fair knowledge in related fields. Assigning a man to several related departments might have been wiser, thus assuring that he would not become the property of a certain department."

The question as phrased suggested that underutilization was a function of Indonesian specialization, but Indonesians who acknowledged the underutilization reversed the charge of specialization: "My experience is just the reverse; IPB expectations were, in a sense, too general and broad! There is indeed some underutilization of abilities, but the pattern of causes is rather peculiar; Kenteam members are expected (by IPB staff members) to volunteer in academic matters, while Kenteam members expect to be asked if so needed!"

In some measure, the problem was another one of failure in communication. The Indonesians were not sure of whether to ask for broader service, so usually they did not ask. This mixture of misunderstandings could have been lessened in its effect by recognition among hosts and guests of the factors involved. Better advance planning for the assignment, including more Kenteam-IPB collaboration in drawing up job specifications, with a proper balance of special and general functions, would have resulted in fuller utilization of Kenteam resources.

Not until 1964 did it occur to IPB to arrange for special end-of-tour seminar meetings with departing Kenteam members, at which professional presentations were made, summarizing or interpreting some feature of development in the field. A simple but excellent idea for the "last exploitation" of a Kenteam specialist, this was arranged only once or twice before "the situation" worsened and the scheme was not repeated regularly. Climactically, however, a paper by one Kenteam member was presented and discussed in a workshop organized by his department during Kenteam's very last week in Bogor.

Ratings of Leadership and Administration

A special feature of the Kenteam self-image is the team's evaluation of its leadership and administration. This is indicated by ratings (grades) given to the three chiefs of party and the three administrative officers.

The ratings of the chiefs of party revealed a mingling of criteria. Some were made on the professional qualities of each man as a specialist in a discipline—agronomy, chemistry, sociology—and some were of leadership performance. Quality of performance was sought in several components of the team leader's task. The first was level of technical (that is, agricultural, scientific, educational) knowledge. In addition, was he a "good representative" of American education, was he an able representative of the University of Kentucky, did he exhibit overall administrative skill, was he a good decision-maker, was he effective in influencing the development of

IPB, did he maintain effective working relationships with IPB officials, with officers in the ministries of the national government, with USAID officials in Djakarta, with the home campus personnel at the University of Kentucky, and with his fellow team members?

The distribution of ratings given by Kenteam members to their chiefs of party are as follows:

Rating	Number of times given
4	1
5	4
6	5
7	8
8	10
9	11
10	8
	47[10]

Ratings for all chiefs of party combined give a composite (average) c.o.p. rating of 7.8 on a scale of 0 to 10.

Level of technical knowledge	8.1
Representative of American education	8.1
Representative of the University of Kentucky	7.6
General administrative skill	7.6
Decision-making	7.9
Influencing development of IPB	7.8
Maintaining effective working relationships:	
with IPB officials	7.9
with ministry officials	8.2
with AID/Djakarta	7.9
with KRF/UKOOP	7.7
with fellow team members	7.1
Overall performance as chief of party	7.8

Apparently most Kenteam members gave their chiefs a "passing grade," but wide variation is noted. It looks as though Mr. Composite Chief of Party at Bogor came out with an average grade of B or B— ! He was best in relationships with

[10] More than thirty-eight because several Kenteam members served with two chiefs of party.

ministry officials, and worst in relationships with fellow team members, but got a grade of no lower than C on any characteristic.

The performance of three administrative officers was also rated by Kenteam members, grading their work in seven task categories. They were evaluated on their success in organizing the Kenteam office and supervising its staff, on their work in connection with the selection and sending of participants, on their effectiveness in processing orders for equipment purchased under the contract, on their expertise in clearing goods through customs. They were rated on their maintenance of team houses and the utilities which served them, on their reception and installation of Kenteam families newly arrived in Indonesia, and on their assistance to Kenteam families who were preparing to leave Indonesia.

The average administrative officer rating for competence and effectiveness, determined on a scale of 0 to 10, was 8.0.

General organization and
    supervising of office staff   8.18
Performance in participant programs   7.70
Processing equipment orders   7.74
Clearing goods through customs   8.49
Maintenance of houses and utility services 8.18
Reception and installation of arriving families 8.00
Assisting families in departure preparation 8.32
Overall performance as administrative officer 8.08

The Fate of Kenteam Recommendations

It was in the nature of Kenteam's work that many ideas for the solution of problems would be generated in daily experience and in the interaction of Kenteam with ipb, the University of Kentucky, and aid. These ideas were routinely expressed in numerous ways as suggestions or recommendations, and many of the most important items of recommendation were written with the annual and terminal reports of team members. Accordingly, a very simple content-analysis was made of the recommendations in these reports, the analysis yields a vignette of this dimension of Kenteam-ipb relationships and to some extent also of the relationship of

Kenteam to AID and to the home university, and of Kenteam's overall effectiveness. This could not, of course, encompass the full range of all suggestions, advisory opinions, proposals, and recommendations that must have been expressed during Kenteam's period of work in Bogor. Nor could it include the recommendations that occurred to and were expressed by Indonesian colleagues. Many recommendations were made fleetingly or deliberately in conversations or meetings, and were ignored, rejected, accepted, or scheduled for further study and consideration. Recommendations which attained written formulation in annual and terminal reports, however, were clearly thought to have special importance.

Eliminating duplication, 335 written recommendations are found in the reports and only two Kenteam members specified no recommendations in their reports. To test whether they were perfunctory and futile or had some influence, each recommendation was checked by some member of the last generation of Kenteam to see whether the recommended action or event had occurred.

Acceptance and implementation occurred by degrees and stages, the typical recommendation facing both modification and delay, but not rejection. Only 10 percent had been followed completely in their original form—just equal to the percentage ignored and rejected.

Accepted and followed completely	10%
Accepted, followed partially or in modified form	41%
Accepted, not yet implemented	19%
Still under consideration, not yet either accepted or rejected	20%
Ignored	3%
Rejected	7%

The fate of 40 percent had not yet been decided—half of them accepted but not yet implemented, hence open to modification, and the other half still under consideration, hence still eligible for either acceptance or rejection.

Most of the recommendations (77%) were aimed at some feature of IPB's activities, organization, or administration, and

75 percent of these had been accepted. Sixteen percent were aimed at an aspect of Kenteam and the University of Kentucky or a related American agency, and 60 percent of these had been accepted. The remaining 7 percent were directed to both or neither IPB-Kenteam, and 54 percent of these had been accepted.

No comparable studies are at hand to show the mortality of recommendations in technical assistance projects, but a 70 percent acceptance rate of major suggestions should not seem discouraging.

Notes in Summary

No standardized criteria have yet been developed for evaluating a group like Kenteam. It is useful but unusual to know something about the self-perceptions of the group and the comparable perceptions of colleagues. The dimensions of the image that we can observe are competence, relationships, attitudes, and strategies. Table I (see Appendix B) contains a sample of items expressing each of the four.

In general, Kenteam gave itself higher marks than it got from Indonesian colleagues, but this must not be taken without two grains of salt. One grain: Kenteam did not make excessive claim to merit. The other grain: IPB abstained from making several judgments, not just to withhold a "low" mark but in doubt or lack of assurance that Indonesians understood Kenteam well enough to indicate opinion. (The shoe was on the other foot when IPB's image was under review, see chapter 8, below.) On the question of competence, however, it must be observed that neither group thought Kenteam's competence was fully exploited while in Bogor.

In competence: Kenteam members had the middle-level technical competence of faculty members in a college at home; in general they were neither the poorest nor the best; certainly one could make a case for sending only the best overseas. But they were lacking on two counts—knowledge of tropical agriculture and skill in communicating their competence to Indonesian colleagues. On competence, the self- and ascribed images were very close.

In attitudes: Kenteam rated itself somewhat lower than in

competence but a little higher than the ɪᴘʙ ascriptions. In restrained commitment to American methods, in understanding of ɪᴘʙ goals, in sense of mission, and in empathic skill, Kenteam gave itself passing marks (that is, half or more attributed these attitudes to most of Kenteam's members). ɪᴘʙ's respondents felt that less than half of Kenteam's members had these attitudes.

In relationship: Kenteam's self-evaluation on relationship items was below the competence ratings but a little above the attitude ratings. They took their places in departments, acknowledging their department heads. For the most part they steered clear of factional disputes. Most had organizational skill and political tact. Their presence in Indonesian meetings was not disruptive. But on each of these indices, ɪᴘʙ's credit to Kenteam was less than the Kenteam self-credit.

In procedure and strategy: These included restraint and indirection, introduction of new methods for teaching and evaluation, traveling for orientation and study. Here Kenteam gave itself one or two high marks and one or two low ones; again, ɪᴘʙ's ascriptions were below the Kenteam self-ratings.

A conclusion which may be inferred from these comparisons is that Kenteam's merit with respect to four sets of characteristics could be marked pass for the team, pass with distinction for some members of the team, and perhaps barely pass for others. It is a little disquieting that ɪᴘʙ ascriptions ran consistently lower than Kenteam's self-ratings on all the characteristics but two—new methods of teaching and new methods of student evaluation.

Kenteam suffered from a quality that was inevitable at the time and that may characterize similar groups in the future unless universities can improve the terms of their participation in cross-cultural, developmental institution-building. This was the quality of compromise seemingly confirmed in the foregoing presentation of images, in which excellence and inferiority of composition and competence were blended in middle ranges. Some of the attitudes which controlled the initial character of Kenteam and its work persisted and some were improved or replaced by more suitable concepts in the later years. Several levels of accomplishments undoubtedly would have been higher if the Indonesian "situation" had been more stimulating to development or if Kenteam's capabilities had been greater, or if both had been the case.

Some of the early and influential views of the Indonesians were that guest professors should be old enough to command respect—if only for age. Prestige and status were in their model of a professor, even though younger men, still without renown in their profession, might have more recent knowledge and more up-to-date methods. One American view was that a generalist was better at the going levels of sophistication than a forward-moving specialist. Another was that one's best men should be kept at home—so let the wanderers do the wandering! Another handicapping view was the fear expressed by some interested but professionally ambitious specialists that they would lose ground in the competition for advancement and recognition if they were overseas and unnoticed for very long. For each such attitude there was also the opposite view—that specialists could help more than generalists, that the best qualified members of a department should get the first call for foreign service, that advancement and promotion would come to the deserving man, whether at home or abroad, that professional growth could be accelerated and not blocked by international experience, and that a contractor university should integrate any overseas operation into its total academic program, staffing it only with regular and highly qualified personnel. It has to be noted, with respect to Kenteam, that only 40 percent of its members were from the home-campus tenured faculty.

8. Images of IPB:
Self and Ascribed

In March 1966, the week after Kenteam professors left Bogor, IPB students occupied the university buildings with sharpened bamboo poles, knives, stones, and kindred weapons, to join the national student demand for dissolution of the Communist party (PKI), reorganization of the cabinet to eliminate communists and to make it smaller, and economic reforms (lowering of prices). The educational program at IPB was abruptly recessed. Nationally, the students had already made some gains in the major cities of several islands with army support; now the pressure was on in Bogor. The main goals were national but some were local. At IPB the rector and his three associates were brought down. A dean and several faculty members were suspended. Events were so urgent that a bystander might think it impossible thereafter to assess the foregoing years of institution-building at Bogor. Certainly the political intensity of that moment took all minds from the lab and the lecture, the office and the library. Recent joint concerns of Kenteam and IPB were momentarily ignored in this new episode in the Indonesian Revolution. Evaluation of the international cooperation at Bogor would indeed be simpler if the surrounding context of struggle could be ignored. But the struggle was real, the complications were part of the adventure, and to develop a college that could surmount them was the task in hand. Lessons of general importance were learned in Bogor's stressful uniqueness.

The Identity of IPB

The symbol of IPB, acknowledging a new university identity, was unveiled in Bogor at a ceremony during Higher Education Week, May 24, 1964. Its basic design is a circle within which there grows a stylized plant. The circle means that science is boundless. The background is blue, the official color of the technical emphasis to distinguish it from social, exact science, and religious institutions. From the design of an open book there arise three stems to indicate the duties—education, research, and public service. From the stems grow stylized leaves which mean that IPB fulfills the five principles of Pantjasila, the national ethic which commits Indonesians to nationalism, humanitarianism, democracy, social justice, and belief in God. Additionally, the five leaves represent the structure of the institute. The two lower leaves refer to the initial two faculties: veterinary medicine and agriculture. The three above indicate that three more were formed at the time of founding. The meaning of the whole picture is growth.

Essential features of the IPB identity have been reviewed in earlier parts of the book but a few additional characteristics require some comment. In particular, two special views of IPB's emerging identity were collected in the survey of Indonesian and American staff members. Continuing the general pattern of reference introduced in the last chapter in the discussion of Kenteam's images, information drawn from the survey is arranged here to see whether IPB's concept of itself with respect to selected traits was identical with or widely divergent from IPB "as Kenteam saw it."[1] The characteristics explored are IPB's manner of decision-making by unanimity, its practice of equal treatment for all, its forward planning, its progress in curricular revision, its acceptance of research, and its use of libraries and books. The characeristics included happen to be items for which both IPB and Kenteam views are a matter of record, and their main significance is in the comparison they afford of the IPB self-image with its image as ascribed by Kenteam.

Consensus and Unanimity

A stated objective of newly independent Indonesia was to develop a "just and prosperous society," and great stress was

[1] Features of the IPB images are discussed here in the order of their presentation in Appendix B, Table II; that is, from the highest to the lowest percentage of IPB members who agreed that IPB "had" the characteristic.

placed on equalitarian values. This normative orientation involved a principle of organization, an ethic of decision-making, and a rule of equal treatment in the distribution of any limited good. The ideological leaders of nation-building effort used the model of the family in their explanations of design. Mutual aid (gotong rojong), decision by consensus (musjawarah and mufaka'at), and executive action by the respected, powerful, and just father (bapak) were promoted as basic principles of organization to be universally applied in political and economic organization, and IPB sought to use them in the development of the institution itself. Kenteam responded ambivalently, appreciating the concepts in ideal form but doubting sometimes the validity and sometimes the efficacy of their application in the development of the institute.

Consensus arrived at by full discussion rather than majority vote was the sanction for action, but IPB learned that the larger a group and the more complex the issue, the more difficult it is to achieve consensual agreement and the more likely it is that two pathologies will develop. At one extreme is indecision and lost time, at the other is the imposition of his view by a strong leader, resulting in what is better recognized as endorsement by the group rather than consensus.

Actually, three ways of decision-making vied for dominance at IPB: the Indonesian commitment to unanimity/consensus, the American/Kenteam tradition of decision by majority vote, and the ever-present possibility of decision by a director-leader or powerful minority. Blends of the three systems were practiced at IPB and tendencies developed to suit decision-making procedure to the type of problem in hand. Decisions of policy called for unanimity or majority agreement; decisions of implementation could be made by responsible leaders.

In viewing itself, IPB saw decision by unanimity as the norm and practice, especially in departments, but almost as much also in the faculty groups and at the overall IPB level. Kenteam didn't see it this way; only a minority believed that the alleged practice of unanimity was "real"; most felt that what passed for consensus was often compliance with a leader-made decision in response to persuasion or imposition. Kenteam was often restless because of what seemed to be indecision at IPB—unanimity was sought but did not exist. There was an interesting difference, however, between IPB and Kenteam responses to the survey question on decision-by-unanimity. IPB responses were yes or no. Most Kenteam members abstained or checked "don't know" and wrote long explanatory comments. One inference from this is that Kenteam members really did not know because they were not very directly involved in decision-making. The Indonesians were clearly operating their own institute throughout the period of Kenteam's presence and many members of Kenteam did not know exactly who was making the decisions or how.

The Principle of Equal Treatment

The survey sought also an indication of whether the principle of equal treatment was dominant in IPB's decision making. Certain Kenteam members had felt that other criteria of priority, such as need or merit, were being overlooked. Did departments get equal treatment in the assignment of guest professors? In the procurement of commodities?

It was felt by a majority of Kenteam members (64%), but less than a majority in IPB (47%) that the principles of "equal treatment for all" and of "sharing whatever is good" were strong factors in apportioning commodity funds within IPB. One Kenteam viewpoint was that this procedure dodged the responsibility of deciding priorities in terms of programs and differential needs. Only 11 percent of Kenteam but almost a majority of IPB respondents (48%) thought that purchasing commodities was too much controlled by Kenteam. Kenteam members were somewhat divided in feeling that they did *not* have (40%) or that they *did* have (50%) as much influence as they wished on the commodity purchasing program.

The IPB and Kenteam views on the distribution of guest professors suggest that the equality principle, while not ignored, was not dominant. The differential needs of departments and the relative differences in importance were taken into account, as were the judgments of the deans, the rector, and the guest professors. The estimates of IPB were in close agreement; on this point, the self- and ascribed images were nearly the same.

Planning

Initially, IPB was short in all the elements for planning, including experience, resources, even planners. Part of the task of development was growth in the practice of planning, and the survey inquired of the IPB and Kenteam samples whether they thought the university had succeeded. The images of IPB as a planner were different. Most of the IPB staff, but only a minority of Kenteam, thought the practice of advanced planning was "well enough developed." The IPB confidence was no doubt a product of the fact that a great deal of planning had been done in recent years, that so many staff members had taken part in so many discussions, and that the visible growth of IPB was in fulfillment of plans that had been made. The Kenteam doubt reflected again the limited participation of the guest professors in what planning did occur and the greater expectations of the Americans based on their own habits of planning, which involved more systematic and continuous attention, and a factor they may have overlooked—working in an atmosphere of available resources, not then possible in Indonesia. One Kenteam member summarized what was probably the general view among the guest professors: "Within the context of a revolutionary society, an extremely difficult political climate, and a deteriorating economy, the planning is about as good as could be expected."

Capacity for Self-Sustaining Growth

A rule of technical assistance is to transfer responsibility as soon as recipients become competent. One problem in applying the rule is that the expert sometimes unconsciously fails to acknowledge the competence of his pupil and holds on too long! A built-in protection for Kenteam was the shift in fields of work from year to year: first an American chemist, then returned participants, and so on through the other areas of work. It was only at the end that transfer was traumatic—to both Kenteam and IPB—because it was mediated not as a normal "disengagement" but under duress that violated the educational norms which would otherwise have prevailed.

In conformance with principle, it was intended from the beginning that the institute would achieve as quickly as possible a capacity for self-sustaining growth. Estimates of how long this would take were not very explicit. The first contract with AID was for a three-year period and it contemplated giving IPB a substantial push. The repeated extensions of the contract, eventually to mid-1966, were recognition by both Indonesians and Americans that more time was required. However, sentiments of nationalism were deepening in Indonesia throughout the Kenteam period and the wish to Indonesianize education became very strong. Ministers and rectors had often to give assurance that they were working toward this end. In 1960 the leadership of IPB announced that all undergraduate instruction would be given by Indonesian staff by 1962 and that most of the need for foreign personnel would end in 1966. The minister of higher education in 1963 was applauded for saying of Kenteam, "Nine years is enough!" The peak of "berdikari" in national expression was reached in 1965, and at Bogor the assertion was heard that IPB had achieved capacity for self-sustaining growth.

A letter from Kenteam to the campus coordinator in February 1965 reported recent steps that had been taken at IPB in a flurry of new confidence in self-maintaining and self-generating capabilities. Provision had been made by the presidium then in office to adopt (by democratic vote and before a stated deadline) a new set of governing rules and regulations, to which any new rector would have to be committed. A plan was announced for democratic nomination and election of candidates for the rectorship. There had been a turnover of institute management with the appointment of new associate rectors, deans, administrative staff, and some department heads. A survey of participants who were studying in the U.S. was being initiated, with Kenteam involvement in distribution of the inquiry but not in treatment of responses (nor was Kenteam informed of results). A complete re-registration of students was made. The demands of an employee-union were rejected. Other measures ordered by the presidium included the making of a complete physical property inventory, establishment of a vehicle pool, establishment of a twenty-four-hour guard of buildings and grounds (under civil defense programs), and adherence to time schedule by the bus bringing employees to work. Orders were given to clean and beautify the grounds. Policies were outlined to promote "upgrading" (advanced study) for staff members within the country by pooling the resources of several institutions, including the universities in Bandung and Djakarta.

It was evident in this package of actions that IPB was tightening the reins of its own self-control and was intentionally removing itself quite rapidly and completely from any dependence on Kenteam, preparing for an early end to the affiliation. The darkening clouds of international tension between the United States and Indonesia were a background for this acceleration toward separation.

In the end, was it agreed that IPB had achieved capacity for self-sustaining growth? The view that IPB could "stand alone" was only variously accepted by IPB and Kenteam staff members, but the latter were less sanguine than the former. In neither group was full recognition given to the length of time required for the complete development of a university. Both groups expressed the view that capacity for self-sustaining growth had been reached "in general"—that IPB could now teach its own teachers. More than 80 percent of each group agreed with this conclusion: IPB agreed fully, Kenteam agreed only in part. The same judgment extended to self-sufficiency in undergraduate teaching. But with respect to graduate teaching, research, and extension work, Kenteam parted company with IPB and did not fully share the latter's confidence that capacity for self-sustaining growth had been achieved. Each Kenteam member answered the question in terms of his own field, and there were many abstentions—those who had served early in the affiliation and didn't know how things stood in later years. The minority of Kenteam who agreed made only a few explanatory comments; most just "checked their ballots."

IPB's claims to self-sufficiency were upheld by reference to the return of participants from studies in the U.S., to the teaching experience of the staff in recent years, to the development of some interest in research and the initiation of some projects, to provision of field and laboratory equipment, to construction of new buildings, and to augmentation of library materials. There were important disclaimers, however, and they support the views of most Kenteam members that full achievement had indeed not been reached—at least not in graduate study, research, and extension. Two illustrative comments from the survey make the point:

Several IPB staff members were well trained, but there aren't enough of them, especially since it is anticipated that this so-called "nucleus" will be depleted by placement of administrators of other institutes, etc. Also the economic situation will have to change drastically before IPB can enjoy appreciable self-sustained growth. In other words, the staff member will have to become a "full timer" and not largely a figure head, as now.

Internally IPB has strength which is not supported by the external society with its political and economic insecurities. These insecurities threaten the life and activity of IPB staff members, so that the potential of the institution cannot be realized. The chief shortcoming of IPB, apart from external insecurity, is that its personnel are trained only up to the master's level. Proportionally there are too many engineers and masters among the teaching staff and too few personnel with training equivalent to a doctor's degree. IPB is now ready for a certain kind of self-sustaining activity, but its growth can only be limited to the extent that it lifts itself by its own bootstraps. It is not fully capable of sustaining growth in higher education.

Curricular Progress

When asked in 1965 whether curriculum development had made enough progress during the period of the affiliation, most IPB staff members agreed "in full" with respect to undergraduate work but only half agreed with respect to graduate teaching. Kenteam was much less sanguine than IPB, with

185

only half agreeing for undergraduate and 17 percent for graduate work.

In responding to this question Kenteam members who served in the earlier years were at a disadvantage. They weren't on hand to know the situation in the later years. Hence Kenteam showed more uncertainty and more dissent than IPB, although there were exceptions.

Kenteam endorsements and demurrers were as follows:

Most effort was on undergraduate level. Graduate program initially was very inadequate.

There have been curriculum committees occasionally but their work has been sporadic and not systematic. Planning within the departments, faculties, and the institute itself has been insufficient. One problem here is that lip service has been given to the idea of having a course in research methods for advanced trainees but this has not been followed systematically.

So far as I know there are no courses (except the recently instituted one on research methods) required for a graduate degree; only self-study and application of method to a particular thesis problem.

Research efforts are weak and badly divided between a major and two minor subjects. Resources and know-how to do research are sorely deficient.

The potential for curriculum development on the graduate level is present—the progress remains to be observed.

IPB has neither the staff nor the facilities for graduate training except for one or two departments.

IPB views inclined toward the brighter side, stressing gains more than deficiencies:

With the help of the committee (with KCT members in it) sufficient progress has been made.

Courses were gradually organized according to overall study objectives and plans. The number of courses and course content have become more reasonable and "up-to-date," although less established than for the undergraduate years.

This has been made possible with the increasing number of successful returned participants in the later period.

Formerly, in forestry, we only took one course in silviculture. Now there are four silviculture courses. No subject has been eliminated due to lack of lecturers. More subjects have been added. Content of courses has been improved.

The progress is that IPB alumni in general know how to handle or to act in certain situations.

Factors Influencing Kenteam Teaching

Kenteam members taught an average of two courses each at IPB, one of the two being newly introduced to the curriculum there. In two-thirds of the courses, the content differed from what these teachers would have used in the U.S. But agricultural colleges in the United States were initially intended for students from farms and their student bodies have always included farm youth. The thinking and teaching of professors in American agricultural colleges has normally assumed farm experience as part of the background of students. For Americans abroad for the first time, it is a surprise to find that students in agricultural colleges are not from farms nor do they intend to become farmers. The usual intention is that they will be employees of the government in some bureau of an agency related to agriculture or serving villagers. At IPB most students lacked any experience or special acquaintance with farming or village life. To see whether this was thought to have created problems for Kenteam, the following question was asked: "Most Indonesian students at IPB have not had previous experience in agriculture and village life, whereas many agricultural college students in the U.S. are from farm homes. This made no difference in the teaching of basic sciences, but did in the teaching of agricultural course content and teaching method. Do you agree?"

A minority of both IPB (39%) and Kenteam members (33%) agreed that lack of student farm experience forced changes in the approach of American teachers. Uncertainty or nonresponse was high among IPB members (23%), and even higher among Kenteam members (39%). Apparently the lack of student agricultural experience was thought to be more significant by the Americans than by the Indonesians, who called attention in their responses to that fact. Kenteam members had no more experience than IPB students in Indonesian agriculture and the teacher's lack of background may have been a more serious limiting factor than that of students in the teaching-learning situation. Indonesian comments, a little more forthrightly than those by Kenteam members, "put the shoe on the other foot"—stressing American ignorance of Indonesian agriculture.

The nature of the Indonesianization process attempted by Kenteam teachers in adjusting the content of their courses at IPB may be illustrated by the several following examples among many which might be cited:

Agricultural Machinery: More emphasis was given to hand and animal tools; pointed toward tropical crops and local conditions, e.g., rice culture, alang [a perennial weed-grass] eradication.

Agricultural Processing: Dealt entirely with estate crops in tropical agriculture: rubber, tea, cocoa, cloves. Rice was also a major concern: cleaning, drying, storing, processing, milling.

Marine Biology: Used tropical and Southeast Asian waters and organisms so far as possible, emphasizing things found there rather than along the U.S. coast.

Taxonomy: The orders, families, and genera of plants native to Indonesia were studied.

Zoonoses: Collected Indonesian incidence data; lectures were slower in pace; all lectures were mimeographed.

Overcrowding with Courses

IPB and Kenteam were asked in 1965 whether or not they considered the curriculum for the first three years to be over-burdened with courses; both groups voted yes. Like all faculties everywhere, IPB had to fight the problem of over-crowding its curricula while still providing broadly enough the education its students required. It was puzzling to Kenteam that so many courses could be listed but it soon became apparent that what Americans might offer in a three-hour course was divided topically and arranged as three courses, each meeting once per week.

The desire to specialize at IPB appeared in the organization of sequences of courses and of courses in new fields. Thus, prerequisites and advanced courses were both made necessary. This did not affect the work in basic science, since the purpose of the chemist, the physicist, and the biologist was to modernize content and teaching method, not to train chemists, physicists, and biologists. At the same time there was the contrary pressure to reduce the number of hours that a student spent in the classroom and to free him for private study and library work with assigned readings.

One factor which must not be disregarded was the desire by many staff members for foreign study and the fear that only one would be sent in each field. This affected planning for future positions. There was too little willingness, under inadvertent or unconscious American stimulation, to select fields of priority and plan to prepare several staff members in a major area. Americans can take a cue from this experience if they wish to give more realistic cooperation in the balanced development of emphasis among disciplines. Another component in the push toward specialization was the prestige anticipated by the only "expert" in any given field. Desire for prestige was a consideration because IPB aspires to preeminence not only in Indonesia but eventually also in Southeast Asia.

A case illustrating these several influences producing overcrowding of the curriculum appeared in a Kenteam professor's terminal report:

At the present time seven separate courses in veterinary public health are taught by full or part-time members of the staff: milk hygiene, meat hygiene, zoonoses, dairy processing, biostatistics, nutrition and tropical medicine, and public health. With the exception of the milk and meat hygiene courses, each is scheduled for one hour of lecture per week during one of the three semesters which constitute the fourth academic year. The planned content of each course is relatively complete and can be adequately covered provided classes are not frequently cancelled. However, it appears to me that the *curriculum drastically fragments the overall topic of veterinary public health.* In brief, I fear that the students perhaps "lose sight of the forest while looking at the trees." Coordination of course content and a better recognition of the importance of proper sequence of topics is absolutely essential to the fulfillment of the veterinary faculty's goal of graduating veterinarians who have a full understanding of the overall purpose and broad philosophy of veterinary public health as well as the detailed scientific information necessary for distinguished service to the public health needs of their country.

IPB was also having or considering a variety of experiences with special training outside the framework of an established curriculum. Short courses were held for managers of estates. In-service conferences were held for field employees of the Agricultural Services (extension). Refresher clinics and short courses were offered for veterinarians who were alumni of IPB. Special instruction was arranged for a few students in microtechnique; it was hoped that this would be repeated, and that a similar kind of special instruction would be arranged in plantology. One recommendation, not yet implemented, was for short courses in poultry. Another was that a one-year training program be offered for IPB graduates who would teach soils at the university level elsewhere in Indonesia.

A problem beginning to be more widely recognized in Indonesia and at IPB was the need for middle-level personnel —technicians as distinct from technologists. Some agricultural high schools and three-year academies had been established to explore ways to meet this need, and it was often a subject of discussion at the agricultural university. It was partly in this connection that various "special" courses were being proposed. It was also partly to provide skill-training in a few fields which were critical but in which only a few persons were needed.

But where in the curricula would be the place for courses in basic sciences, in economics, in sociology, in biochemistry, in home economics, in genetics, in food technology, in administration of agricultural agencies, in extension and public service, in engineering? There were "tugs of war" on all these points. Logic led some thinking, even in the early years, to suggest a science college, additional to the others and specializing in the teaching of all courses for the first-year and probably the second-year students in all colleges. (It was ironic that the University of Kentucky, parent of Kenteam, itself undertook this fundamental reorganization on its own campus only in Kenteam's last year at Bogor!)

The nature of the problem was seen in animal agriculture (the College of Animal Husbandry) in which it was agreed that courses should be included in soil conservation, agronomy, pasture and forage crops, transportation, food technology, technological areas other than food, disease control and prevention, sanitation, farm management, and agricultural engineering, as well as breeding, reproduction, nutrition, management, and production practices. On this list of fifteen items, at least nine would normally be taught elsewhere in the university than in the College of Animal Husbandry itself.

There were alternatives and each was followed. One was to establish a department in one of the colleges and commission it to serve all the other colleges. Another was to establish a department in each of two or more colleges and let duplication lay its chips in two or more piles. A modification of these plans was to establish a new specialized college.

The Fifth Year

The fifth year at IPB was initially puzzling to Kenteam members, who were familiar with the traditional four-year college period in the U.S. At first there were curriculum suggestions which would introduce a degree or certificate after four years, representing a major shift from European to American practice. The attention of Kenteam turned from this as a major immediate objective, however, and curriculum questions focused on general preparation in the first years and some concentration on field or "practical" experience in the fifth year to complete requirements for the Ir. degree or the veterinary doctor degree. Probably some misunderstanding arose from the tendency of both Indonesians and Americans to translate "prakticum" too literally as meaning practical, whereas the connotations of "laboratory" were an important part of the prakticum concept. Toward the end of the Kenteam period at Bogor, the expectations for the fifth-year prakticum (in fact, a prakticum in any year) were centering more on researchlike effort involving practice of skills to some extent, but controlled observation as well. The plan for the fifth year in the fisheries curriculum is an example. The student does a comprehensive research problem and writes a thesis. In addition, a library research report (scripsi) is required. The topic for the scripsi is approved by the major professor and normally relates in some way to the student's major subject. Then, before graduation, each student must make a presentation in a seminar to the staff and students of the faculty. Finally, the student must successfully complete an oral examination which is presided over by a five-man committee. Opinions about and understanding of this fifth-year procedure varied among IPB staff and Kenteam members. Half of IPB but only one-fourth of Kenteam agreed that the fifth year *sufficiently* stressed the aims of discovery, in contrast to the stress on learning by memory in the earlier years.

Books and Library Facilities

None of its developmental problems was more urgent for IPB than getting books and promoting their use. Only the professors had really needed books under the traditional methods of instruction and the students in universities had relied mainly on lectures. But the desperate scarcity of books at IPB, like the nonexistence of teaching laboratories, retarded progress in every aspect of the institution's life. The physical presence of books anywhere expresses the culture which produces and uses them, but in Indonesia even the Koran was learned by ear and the masses could not read. But the campaign against illiteracy and the spread of elementary school work were moving like wildfire in the early years of Indonesian independence, and the production and use of books would someday surely follow.

At IPB—and in Indonesia generally—there was a void of the resources from which books could come: no demand from a student body of trained readers, no paper, hardly any presses, and who would be the authors? Neither the culture in general nor its systems of education had yet required the use of books, so there were none.

Textbooks, books for supplementary reading, reference volumes, scientific journals bringing in reports of research to update knowledge—these were part of the habitual and daily instructional material of American teachers. With their IPB colleagues they had to begin immediately to seek books—relying, of course, on American sources. A gesture of goodwill by students at the University of Kentucky brought forty gift cases of used books from Lexington. These were a miscellany of general literature, received with appreciation, housed inaccessibly, and used only a little.

The first orders for purchase of equipment with contract funds included many for books. Apparently, even though previously without books, beginners in their use, especially students, have an instinct to seek what is new and up-to-date. They want to enter the house of modern knowledge through the front door. In fact, even before Kenteam's arrival, AID had ordered a beginning supply of background books in agriculture and a small printing press for IPB. Some Kenteam members assembled large collections of bulletins from American colleges and informational publications from American commercial firms. Students could not afford to buy books even if available. It was not feasible to provide a text in each course for each student, so procedures of loaning and renting were tried, and sets of books were shelved where space could be found; this was usually where the departments were housed and where rooms could be set aside as libraries. But the book-hunger of professors was greater than that of students and it was impossible to keep volumes from getting into offices and homes where the instructors could really use them to best advantage. Books disappear like moisture in a desert until there is enough accumulation to stock central places of common supply. The main purpose of getting books into the hands of students had to wait for fulfillment until part of the appetite of the professors was fed. And, in fact, the whole purpose was never fully accomplished or even well enough started in all the time Kenteam was at Bogor.

Books needed to be available in the Indonesian language, not just in English, Dutch, German, Russian, French, and other languages. Regardless of politics, however, English has become a world language in science. It was not a bad thing that Kenteam brought American books to IPB—it was just not enough! The provision of paper, ink, presses, and the reward of authors for producing volumes in Indonesian were still among IPB's great and unsolved problems when Kenteam went home. The production of Indonesian texts and laboratory manuals in a few fields *did* get started, however, by a few instances of collaboration between Kenteam and IPB colleagues, and several returning participants had ambitions to write books and articles in their fields.

Selection, acquisition, cataloging, and controlling the use of books require libraries and librarians. There need to be libraries in a system of departmental, college, and university central locations; libraries must be open for student use; instructional methods and methods of evaluating student achievement must insist on the use of books and libraries. A decree by the minister of higher education in 1963 set the stage by calling for a central library at each national university.

The idea of a central library was accepted as "logical" at IPB, but practical considerations resisted its implementation. The distance of dormitories from any central spot and the wide scatter of college buildings were elements of resistance. Library organization at IPB and other Indonesian universities remained seriously underdeveloped.

A majority of Kenteam members (56%) observed that in the development of library facilities and in the use of books and journals generally, IPB had made great progress during the affiliation. There were many who did not comment; it is clear that neither the first team members nor the last in Bogor could have seen the situation at both the beginning and the end of the contract period. IPB staff were there at both times, however; hence, only a few failed to respond to the question. A large majority (90%) agreed that there had been great improvement.

All the newest features of information storage and retrieval will need to be examined for prospective uses in the developing countries. With microfilm and modern systems of bibliographic memory and consultation, universities and libraries need not follow laboriously the steps by which the traditional libraries have evolved, only reproducing outmoded systems, when ways may be found for them to enter the twenty-first century as quickly as anyone else. But this is further than IPB and Kenteam were able to go together during their affiliation.

Status Report, 1966

Much had been started and nothing finished in the quarter of a century of institution-building at IPB which ended with Kenteam's last year at Bogor. Founded by the Dutch in 1940, suspended during the Japanese occupation, reestablished in the zeal of independent nation-building, picked up and moved along with new technical assistance from abroad, the Agricultural University at Bogor had struggled through a maze of restraints and deficiences, had achieved autonomy and multi-college complexity, and was charting directions of needed further development. Against the political turbulence in the background and the unstable context of economic support, IPB had established its identity as the dominant institution of higher agricultural education in Indonesia. Through the University of Kentucky team it had drawn heavily on American experience and American supporting assistance, but was achieving unique Indonesian institutional character. It was poised against threatening collapse, and with a frame of organization and structure had some short-run capability to ride through hurricanes without permanent deflection from course—if the storms didn't last too long!

Indonesian and American evaluations of the university's identity were as different as their characterizations of Kenteam. Kenteam regarded itself a little more favorably than IPB regarded Kenteam. Similarly, IPB rated its own qualities a little higher than Kenteam rated them. Each identity—that of Kenteam and that of IPB—reflected its own self-confidence, some of its own self-pride, and the application of its own system of values. Correcting for these two sources of bias, the evaluations—each group of itself and each of the other— were not widely different.

With expanded student and staff personnel, IPB had partial autonomy in relation to the ministries and other institutions. It had internal structure of sections, departments, and colleges, and a developing curriculum; it was improving its system of decision-making and its criteria for decision. It had gained strength through acceptance and use of American aid, mediated by Kenteam. It had acquired basic facilities for teaching and more were in sight. It was being linked to ministries, services, local governments, organizations, estates, and farms.

The Agricultural University in Bogor in 1966, by self-effort, by Kenteam's nurture, and by some aid from other sources, was an emergent institution, but it was a mistake for anyone to think that IPB had achieved full capacity for self-regeneration and maintenance of thrust. With all the jerkiness of forward movement, reversals, swerves to left and right that IPB had felt, twenty-five years could not have been long enough to build the institution and the long-time fate of IPB could not, in 1966, have been predicted. Kenteam and IPB had made good use of their time together, but their task was not finished when the affiliation was broken.

9. The American Response to Indonesian Culture

The Overseas Wife

Comments on the selection of personnel to work abroad usually stress the importance of screening wives as rigorously as men. Much is made of the high visibility of American women in foreign communities and the prominence of a woman's role in affecting the tone of relationships and in augmenting or detracting from the effectiveness of whatever work is being undertaken. For all the homage to this idea by employers of men for foreign assignments, however, there is little evidence that more than casual attention is given to the qualifications of wives. One is tempted to conclude that if the matter were as important as is argued, it would get more careful and systematic attention. Generalizations about American wives overseas are mainly impressionistic and studies are needed of their roles and competences.

Bogor was viewed as home for at least the next two years by each Kentucky team family on arrival. The prospect was potentially exciting for any newcomer; it was met with eager anticipation by many, with trepidation by some, perhaps with resignation by a few. The response of the American families to the adventure of life in Indonesia was a prominent variable in the performance-formula of Kenteam. When a man's family is with him at the site of overseas work, it may be assumed that "as the family goes, so goes its head."

Thirty-six Kentucky team wives lived in Bogor; data are included in this chapter from twenty-seven of them who filled in special questionnaires about their reactions to Indonesian culture. The questionnaires were formulated on the basis of preliminary discussion and later pretesting with Kenteam wives who were in Bogor in mid-1965. The questionnaires were then mailed in November and December 1965 to all Kenteam wives who had ever served in Bogor. The twenty-seven who responded are an 80-percent sample of the total number to whom questionnaires were sent. The wives who responded ranged evenly among the working ages: none were under thirty and only one was over sixty. Ten of them went to Bogor from Kentucky.

Motivation

The stated motivations of Kenwives for agreeing to work and live in Indonesia confirm their sense of mission. Their explanatory comments make it evident that this was a dominant, though not their only, motivation. Considerations of self-interest and status were present also, but not to the extent of concealing the altruistic tinge. The mixture of their motives would seem in general to have qualified Kenwives for life in Bogor. They listed reasons for accepting the Bogor assignment in the same priority as their husbands: first, to help with an educational project; second, to travel; then, in further order of mention, for a bigger salary; for a change in job and family living conditions; and to get ahead.

The distinctive motivation of Kenwives—and in fact of Kenteam, too—was an occasional matter of comment by Indonesians in Bogor, who distinguished the educator-Americans from business-Americans. The former, they thought, were more interested than the latter in learning about Indonesian culture and in associating with Indonesians. They felt the latter were content with "goodwill," which was something less than friendship.

Preparation

Preparation for the work and life in Indonesia was a pre-occupation of Kenwives during an interval that many of them thought was too short. The decision to accept the work in Bogor was made an average of six months before departure, but a few had less than two months in which to get ready. Several of them recalled preparing hurriedly and with a sense of pressure because of limited time. Their problems in getting ready, were, in order of mention: studying about Indonesia, deciding what to take, borrowing money for expenses, planning school for children, and arranging for care of the house. Difficulties in arranging for a leave of absence and making arrangements for family members left behind were mentioned only a few times.

Borrowing money was a major concern. More than half reported they had to borrow in order to finance preparations and travel. One family borrowed $600 from a bank; another borrowed from parents; one family spent $2,000 of savings and borrowed $500 more; two families had saved to buy new cars and used these savings; one family borrowed more than needed to buy linens and clothing for two years ahead; one borrowed to keep from using savings. Apparently most of what was borrowed was repaid from salary during the first year overseas, but the use of credit was a fairly common device in meeting the costs of preparation.

Deciding what to take to Bogor was a family preoccupation during the interval between decision and departure. Some of the wives found it to be no problem; they liked shopping and enjoyed the special event of shopping for two years ahead. Many later concluded that they had come overprepared, that their advice from others on what to bring had been either erroneous, incomplete, entirely misleading, or contradictory. Such statements, however, merely reveal differences between the families and the fact that they didn't all need the same things. In general, the wives who felt that deciding what to take was a problem were inexperienced in living abroad and lacked the confidence with which they might have approached a second experience.

Problems in arranging for care of the house were reported by about half the wives and half reported that planning schooling for children was hard to complete in the time between decision and departure. Solutions to the latter were varied; some took Calvert graded school materials with them; some took regular texts that were to be used in the home school during the period of absence or books suggested by the teacher in Bogor. (American parents maintained a cooperative tutoring school—the "West Java Day School"—for their elementary grade children in Bogor, sharing costs and finding teachers from within the group.)

In retrospect, Kenwives offered suggestions for lessening

the problems of preparation. There was almost unanimous agreement on two themes: language and cultural orientation. Kenwives regretted that they hadn't been taught Indonesian. The usual suggestion was for a short-intensive course and continued study after arrival. They felt also that they should have been—but were not—given intensive and realistic instruction in the history and culture of Indonesia.

Areas of Shock and Frustration

Given appropriate motivation, a commitment to the enterprise, and the fulfillment of preparatory activity, what happens to an American wife during two or more years in a strange host society? The cultural shock idea is frequently used either for comparative reference or for attempted explanation in answer to such a question. Most or all of the Kenteam families were thought by their colleagues to have overcome cultural shock and to have become adjusted by the end of their first year.

Of course, the life of an academic wife in the United States is not without its problems. To study the frustrations of Kenwives in Bogor is not to imply that there would have been none at home or that life overseas was just a tangle of troubles. But the objects and situational contexts of both frustration and satisfaction are different abroad. It is common knowledge that pressures of housework in the U.S. are vexing in spite of gadgetry; that there are competitive struggles for social status, problems of child development and school, role conflicts; that pressures on time and money are sometimes overwhelming; that desires are often unfulfilled; and that family and community relationships can become snarled.

But cultural shock is a nonspecific, spongy concept. To get sharper focus, attention must turn more concretely to the objects, the degree, and the duration of frustrations denoted by such terms as alarm, anxiety, apprehension, bafflement, depression, discomfort, embarrassment, disturbance, impatience, inconvenience, irritation.

Participant observation in Bogor and pretesting in the last group to work there resulted in the identification and listing of eighteen areas of frustration in the experience of Kenwives, and the questionnaires asked about each area. Each wife graded the strength of her feeling during the first few weeks after arrival in Bogor, after six months, after one year, and at the time of departure from Bogor. The review of Kenwives' experience has meaning beyond its face value. What was happening to Kenwives was very central in the whole experience of Kenteam, and one may be sure that the major frustrations reported by the wives were largely shared by all the members of Kenteam families.

The areas in which Kenwives experienced frustrations and the number of wives who indicated each are as follows: news (24), traffic (23), poverty (20), health (20), language (19), standards of living, especially housing (19), cleanliness and sanitation (17), time schedule (16), etiquette in daily life (14), population density and growth (14), atmosphere and slogans of the Indonesian revolution (14), values and beliefs (14), money (12), food (11), and Indonesian culture in general (6).[1]

Response to Indonesian Culture in General

Of the eighteen areas of experience, Kenwives said the last, Indonesian culture in general, evoked the least frustration. Only six of them reported responding with feelings of anxiety and frustration. The reasons are revealed by information on the activities of the women and the generally pleasant participation of most in many facets of Indonesian culture. It was only in the cases of inactivity and restricted participation that unconquered frustrations endured beyond their normal life. Maximum participation produced minimum frustration.

Friendships. An indication of the range and depth of their participation in the Bogor experience may be observed in Kenwives' reports of their friendship with other women.

[1] Except for "Indonesian culture in general," areas of frustration are discussed here in the order of their presentation in Appendix B, Table III; that is, from the largest to the smallest number of Kenwives reporting feelings of frustration after six months in Bogor.

Friendships were graded as close (intimate), neighborly (involving reciprocal visiting), and casual (one-way visiting, or "you greet when you meet").

Kenwives reported having an average of 6.1 close friends among women in Bogor, 23.1 neighborly friends, and 61.8 casual friends. The average Kenteam wife had 11.5 women friends within her own circle of Kenteam wives: 2.2 close, 7.4 neighborly, and 1.9 casual. The maximum number possible, of course, was limited by the size of Kenteam and the duration and overlap of Kenteam members in periods of residence in Bogor. The average Kenteam wife had 62 Indonesian women friends, 2.5 close, 9.0 neighborly, and 50.5 casual.

There were a few other possibilities in Bogor: foreign women from other countries than the U.S. whose husbands were also engaged in some technical assistance activity in Indonesia, or other American women whose husbands were associated with business enterprises. Among these others, the average Kenteam wife had 17.2 friends: 1.3 close, 6.4 neighborly, 9.5 casual.

In friendships, Kenwives formed their judgments of each other and of Indonesians. Asked what they thought of Indonesians, the Kenwives' dominant theme was strong affection. "I love them," commented several. Indonesian traits and characteristics as Kenwives saw them were intelligence, tolerance, happiness, kindness, beauty, weakness, talent, patient gentleness, an ability to adjust, artistic talent, sensitivity, naivete, cleverness, peacefulness, pride, sincerity, pride in country, a desire for education, and a craving for understanding. Nearly all of the Kenwives contributed to the list of descriptive terms. There were a few negative assessments: their lack of initiative after 350 years of colonial subjugation, lack of leadership, the precariousness of their economic lot, "color complex," "mixed-up country," lack of experience and knowledge, fear.

The feelings expressed by Kenwives about Kenteam families in general were of two kinds. One was a quality judgment offered by most of them to the effect that Kenteam families "on the whole" or "for the most part" were a "fine" group, a "dedicated" group, but with a few exceptions: some "should

not have been in that environment" or were "badly chosen."
The majority of Kenwives clearly thought that a majority, but
not all, of Kenteam families merited a positive rating. The
other reaction was one of friendship and affection, again
expressed by a majority for a majority, and again with exceptions. They were "very fine friends." "Wish I could see
more of them now." "We hope to keep in touch." But some
"needed more training in sharing." The general tone was
that these wives were such a cross-section as one would find
in an American community. There were the socially conscious,
the selfish, the ambitious, the missionary-minded, the genuine,
the bridge-players, the tukang (peddler) buyers, and the plain
ordinary housewives.

Activities. The wives of American university professors in
their home communities are known to engage in many family
and community activities. It is to be expected that they will
seek similar interests, if possible, when living abroad. They
take up residence in a new community free of all former
obligations outside the home, and with servants who rearrange
their use of time in the household. Their lives are thus freed
for new programs.

A fairly complete listing is available of the activities of
Kenwives in Bogor, showing great variation. Some of these
activities were organized projects and some were individual
expressions. They involved different combinations of Indonesian and American persons and cultures and different numbers of persons, sizes of groups, and degrees of cultural
interpenetration.

Many of the activities in which most wives took part often
were the organized-project type: church, women's association,
language study, English teaching, supporting the orphanage.
Activities often or occasionally engaged in included matters
of shopping, etiquette, and friendship. Bargaining was high
on the list, with peddlers at the door, or in shopping at the
local market for consumption items, batiks, and wood carvings.
Among gestures of etiquette and friendship were sending
flowers and Lebaran (New Year) greetings, and entertaining
friends.

The activities in which most Kenteam women engaged

rarely or never were mainly personal hobbies. While none of these activities attracted a large number of wives, however, each was of interest to at least one. A few did report playing golf and tennis, keeping diaries, playing bridge with Indonesian and American friends, expressing themselves musically by piano or voice, learning anklung or gamelan music, the Sundanese language, or regional dances.

Teaching. One activity reported by most of the wives was teaching English to prospective participants, to wives of IPB staff members, and to others in the Bogor community. In the first years of the affiliation, wives carried on this work as a semiofficial task in preparation of participants for their study in the U.S. They were especially interested in this because it gave them an important role in Kenteam's mission. In the later years, this work was done by Indonesians who were employed for the purpose by AID, and the wives shifted their language teaching to others who wanted to learn, continuing, however, to sponsor informal discussion and conversation practice meetings for the participants. They taught English at the Nutrition Academy and Bogor Teachers' College, to high school students, faculty members, IPB employees and students, and Bogor police and their wives.

It is important also to recall the part-time employment of a few Kenwives as teachers of the American children in the locally sponsored tutoring group. (This could not be established formally as a school because of an Indonesian government ban on foreign schools.) Each year this work was organized by an education committee of American parents in Bogor and supported on a per pupil basis by the parents whose children were enrolled. In retrospect, the Kenwives who served as teachers of American children expressed one note of regret because the time involved kept them from participating more in other activities. They did not fully enjoy working and felt they "had missed something."

Cultural Participation. Most Kenteam women reported seeing or attending several kinds of ceremonies and performances of drama, dance, and music. Nearly all had attended weddings, special selametans (socio-religious ceremonies of communion), wayang (stylized native epic drama in shadow play, puppet show, or on the stage), gamelan (native percussion orchestras), and dancing (Javanese, Balinese, Sumatran). Some of the rites of passage (birthdays, baby hair cuttings, circumcisions, funerals) had been attended by half or more of the women. Those seen by nearly all Kenteam were relatively public in character, for all interested spectators, and they were frequently programmed by Indonesian friends especially for the Americans. Those seen by fewer (although still by a majority) were more intimately restricted or private to family and friends, and their meaning was more religious than dramatic or recreational. It is clear, however, that Kenteam wives had been introduced to a variety of artistic samples from the culture in which they were guests.

Nearly all wives had been in Indonesian homes for tea and as dinner guests, but only five had ever been overnight guests there. In the routine of life in Bogor, of course, there was no occasion to be overnight guests; this role could have been experienced only in case of travel.

The activities of a group of temporary residential guests in a host community may be largely internal to the guest group and insulated from the surrounding host society. At another extreme, they may as persons become merged into the host society, reflecting full assimilation. Neither extreme is likely or normal. In the case of Kenteam, the first was unlikely because everyone had already made decisions to be overseas and to work in the host country, thus expressing or at least implying an interest in the host society and culture.

Activities That Meant the Most. Kenwives expressed sentiments of appreciation and enjoyment of their participation in Indonesian culture.

I enjoyed playing and singing the Indonesian songs and having songfests when Indonesians visited us, especially the students. I acquired many examples of Indonesian arts and crafts and spent much time studying their production.

I enjoyed visiting and entertaining Indonesian friends. When one is away from one's own family it is satisfying to develop a family situation with the local people of similar interests.

The warm friendships with the women of my English study groups continued. Being invited to be a member of the 60-piece anklung orchestra called Saraswati was one of my most precious experiences. Meeting twice weekly for practice sessions was fun. These were "ibu-ibu Bogor" [Bogor mothers], ordinary housewives, a very different stratum from the faculty wives and women in my English groups.

I enjoyed the Bogor-Indonesian women's meeting—made many good friends, learned much of their customs, etc. I visited in homes and was told this group had done much to help the Indonesians get a better understanding of the Americans and their way of life.

The last two years I was in Indonesia I became very interested in Indonesian music and learned to play the gender. My teacher taught me a lot about Javanese gamelan and I learned the Javanese songs (singer's part) as well as the saron and gender (instrumental) parts. My last year in Indonesia I studied Sundanese music, learned to play the ketjapi, and also studied Sundanese singing.

I especially enjoyed Chinese cooking lessons and a weekly conversation group of Chinese Indonesians. The forestry-staff wives had weekly meetings in their homes and special events in my home—very friendly.

I enjoyed going to all the ceremonies because my husband was able to photograph these in order to tell Americans about them. The most worthwhile thing I did was to prepare participants for American ways of life—using the English language and explaining our customs.

Of their anecdotal comments on activities that "meant the most," all but two dealt with a joint American-Indonesian event or relationship.

The Kenwives' evaluation of their own "different" life in Bogor was expressed by most of them in superlatives of enjoyment and appreciation. There were two or three exceptions —wives who, in retrospect, felt that they had not enjoyed the experience. The most-often-noted feeling was that residence in Bogor was a "high point in my life": "happiest time," "enjoyed every minute," "privilege," "treasure forever," "fondest moments." Other expressions were of the same character and nearly all were effusive. Kenteam wives, without exception, said their experience in Bogor was "worth it." There were none who did not describe some gain, and seventeen said they lost nothing. The reported losses were weight (mentioned by three), health (mentioned by two), time with family and friends left behind (mentioned by three), material objects, cultural advantages (such as good classical music), and "some preconceived notions." Even those reporting losses said the gains were greater.

The gains were friends, cross-cultural knowledge, a broader view of the world, new technical interest, deeper involvement with foreign students at home, memories of travel around the world, better self-understanding, self-assurance, a change in values, greater appreciation of the U.S., a more frightening insight into world problems, respect for the resourcefulness of the "have not" people, and appreciation of the simplicity of living.

"The gentleness, friendliness, and ability to enjoy life with a minimum of material possessions shown by the Indonesian people made a profound impression on me." "Things I saw and heard were better than another degree from any college." "I came to the relaxing recognition that other people and ways may be different, but are neither necessarily better nor worse than the American way. One must judge (if at all) in the context of the situation." "I achieved a far bigger outlook on the rest of the world. I know that we must help people to help themselves. I overcame my race discrimination, too."

Regrets. When asked to recall what they wished they or others had done differently, three-fourths of the Kenwives had answers. Nearly all their wishes were that they had done more to learn the host language and culture, that they had travelled more in Indonesia, spent more time with the Indonesians and less with Americans, lived more simply, spent less time and money shopping, taken less clothing, food, and supplies from home, been less critical, given less emphasis to "the way *we* do things," learned more about Indonesian culture, had less "worry and fret," and that Kenteam members had been less patronizing. These expressions of regret all concerned human relationships and not things, and revealed a fairly general wish among Kenteam wives that they had been more deeply involved in learning about and experiencing the host culture.

Areas of Frustration

News. The foregoing discussion has dealt with culture in general, which was not an identifiable area of frustration. But the other seventeen areas of experience were a different story. The most prominent subject of anxiety among Ken-wives, and thus by inference among all the Kenteam, was the absence of news. Twenty-two of the wives reported initial concern, and the number was even larger (twenty-three) at the time of departure.

Americans at home live in a saturation of news in daily papers, radio, television, and telephone conversation. Attention to news varies from indifference—letting the world go by—to intense interest and keeping up to date. Reduction in exposure to news is a major deprivation to Americans abroad. Most local newspapers and radio broadcasts are in a language not read or understood. If there are local newspapers in English, they have limited coverage and may have political bias. Newspapers and magazines from home come late and irregularly. Shortwave radio and letters from home take on major significance, and mail deliveries are eagerly watched.

Nearly everyone in Kenteam read an English-language newspaper. During most of the period at least one and usually two were published in Djakarta. But the slant of the news alternately angered and amused Kenteam readers. Reading newspapers in Indonesian was impossible for most of the team—only two reported trying it regularly. All Kentucky team homes received American newspapers and magazines— but late and irregularly except via APO in the later years. For a time, some Kenteam members had access to the European edition of the *New York Times*. Many popular news magazines were banned during parts of the period when Kenteam was in Bogor. "All we wanted were banned—the others never arrived." Summaries of major news items were mimeographed by the American embassy in Djakarta and were a much-appreciated source of international and some domestic U.S. news for Americans in Djakarta and Bogor.

Shortwave radio was actually the chief source of news, and a variety of broadcasts were available. The Voice of America and Radio Australia were heard by thirteen families each; Radio Mayala/Malaysia, the British Broadcasting Corporation, and Manila's Far Eastern Station were heard often; one wife reported monitoring English-language broadcasts from Radio Moscow and four from Radio Peking occasionally; three listened to Radio Djakarta, which offered occasional broadcasts in English. But English newscasts were frequently jammed.

Telephone lines and equipment in Bogor were old, badly worn, and replacement parts were not available. There were telephones at IPB and in two or three of the Kenteam houses, but their operation was hesitant and sporadic. To some Kenteam wives, "it was a pleasure to be away from the phone." Much local communication was by notes which were sent with the family driver or another servant and "the note-sending system worked fairly well." Mail service was of very great interest and importance to Kenteam, and the wives expressed their reactions in a variety of comments. There was lack of confidence in local and international mail; in the early years of Kenteam "the pouch" (embassy facility) was used for some mail to the U.S. In the later years, the APO service was "a huge morale booster."

Not everyone was bothered by the lack of news, however. One Kenwife commented, "I was glad we weren't getting all the news about Indonesia when we were there. It was alarming in the U.S. Our families were sometimes frantic—but we, living there, never realized this."

Traffic. The second most frustrating area of experience for Kenteam families was traffic. The systematic control of traffic has been so regularized in the experience of Americans that unregulated and unpredictable movements provoke real fear. The movement of traffic in Indonesia is controlled more by what vehicles will do and by the impulses of their operators than by laws and officers, although both of the latter exist.

The lanes of travel were well worn before the day of automobiles and they are still considered by the masses to be the property of the pedestrian, the bullock, the pony, and the bicycle. Keeping to the left instead of the right, speeding in slow places, operating unfit machines (some of the vehicles used in the 1950s and 1960s had been buried during the Japanese occupation, then disinterred and restored to use later), the mingling of man-powered, horse-powered and engine-powered units in unbridled and full-throttled competition for the road, the supremely confident driver and the oblivious street-crosser—these were the traffic of Indonesia as Americans saw it and to which they responded with mingled fascination and fear. Kenwives went somewhere almost daily by car or otherwise through streets and over roads, and their frustration lay mainly in the unpredictability of vehicles and their drivers. Most procured Indonesian driving licenses and drove occasionally (each Kenteam family had its own car). However, most families employed Indonesian drivers for convenience because they knew the places to go and this prevented much language frustration. Only five of the Kenteam wives tried cycling, four traveled by oplet (a small bus) and two by bemo (a motorized triksha), three traveled by bus and three by train. These, with two additions, were the vehicles used most commonly by Indonesians. The others were the betjak (a tricycle ricksha) and the delman (a horse-drawn buggy); nearly all Kenteam wives had tried them, but none used them regularly. It is clear that all regular Kenteam travel on Java was by car; interisland travel was by boat or plane.

All Kenwives were disturbed by traffic on arrival in Indonesia; only four had overcome their anxiety about traffic before departure from Bogor, but about half of them had taken some kind of action to reduce their frustration. All but three had hired drivers. Only six had curtailed their own travel through fear or dislike of traffic. There were many comments on the traffic question, reflecting the extent of concerned interest.

I had difficulty getting used to driving on the left side.

Our driver tended to play "chicken" with every approaching car;

I finally adopted a fatalistic attitude, relaxed, and stopped watching the road.

They took chances I never would have taken, such as passing on blind curves, driving too fast.

Even with a driver my anxiety never lessened. Every time we were in the car, I was tense and worried.

After the first year, my desire to see and travel was stronger than my fears.

I had no particular problem other than being scared to death a number of times.

It was a rather shattering experience to any American. Most Indonesians drive cars at night with only parking lights on—no lights until opposite you, then they turn them on full. Thus it's a good idea to use a driver on the highway at night. I did all the daytime driving on my own—enjoyed the challenge of crowded streets, a new experience.

I eventually convinced myself that betjaks [trikshas], sepedas [bicycles], and pedestrians were under some form of special protection and my concern was not needed.

Eight of the wives reported accidents; six were automobile collisions in which Kenteam family cars were reportedly hit by other cars; there were two cases of dangerous electric short circuits caused by flooding in heavy storms, and one bicycle wreck.

Poverty. The economic problems of a country struggling to develop appear in dramatic prominence to visitors from societies which have reached other stages, and Kenteam went to Bogor before the visibility of urban poverty in the United States had startled the middle-class population back home. Inflation, with rising prices and continuously low wages, deepens poverty, with its hunger, undernourishment, unemployment, ill health, bad housing, lack of clothing. The impact of visible, dire, and chronic poverty is severe on middle-class Americans who come to live and work in technical assistance in developing countries. Poverty was the third most prominent object of depression and concern. Nearly all Kenwives were strongly frustrated by it and just as many were frustrated at the end of their stay as at the beginning. Several of them attempted some kind of alleviative action; others just "got used to the problem." All the Kenwives did give some kind of help to somebody: food, clothing, loans, medicine.

Malnutrition for many and undernourishment for nearly everyone, I never could accept without guilt.

Our concept and that of an Indonesian concerning poverty cannot be compared adequately—we just do not have the same basis of comparison.

I don't think I could ever get used to this; but you must do what you can and not become morbid over a situation you cannot correct.

One never gets used to suffering. We arranged for the orphanage to have eggs with their fried rice on Sunday mornings. We gave them an occasional sheep after we learned that normally they had meat only at Lebaran.

Rather than make outright gifts, except at Christmas and Lebaran, we made jobs for which they got extra pay. We tried to keep the wage scale consistent with local practices.

We made loans (but few gifts) and tried to help them solve their problems—a gift would be only a temporary solution.

I helped women learn the use of corn to stretch their diet. I encouraged crafts, home projects to increase income, making over clothes, and better ways of cooking to preserve food value.

What worried me most was what happens to these servants whom we teach when we go home?

I felt guilty particularly when I sent cook to market with 50 or 100 rupiah for food for one day—almost more than he made in a week.

I felt guilty that I had so much to eat and friends had so little; I was particularly concerned about children. We gave away quite a bit of our canned food.

We gave powered milk for children of friends, and gave our servants medicine, clothes, 'rupes.'

I saved everything I didn't need for somebody and gave it as needed—even jam jars, empty powder boxes, socks, scraps of cloth, elastic, etc.

Health. Threats to health are numerous when Americans go abroad to live and work. Indigenous death rates remain high in many areas although they are coming down slowly, while birth rates remain high and population increases. In Indonesia, dysenteries are almost certain, the risks of contagion and infection are high, and medical and paramedical personnel are scarce. Exposures to amoeba and parasites are common, malaria and tuberculosis are widespread, and many Americans don't stay long enough to build up any of the resistances prevalent among indigenous residents. In addition there is the worry for friends and associates among whom the ravages of illness and nutritional deficiences are in constant view.

Health problems were an object of anxiety and frustration to most Kenwives at the time of arrival in Bogor; there was some decline of intensity by the time of departure, but anxiety and frustration remained for many (eighteen). Most of the women (nineteen) reported taking some action to overcome their anxiety by caring for the health of their family or by providing medicines or treatment for friends. They weren't trying to solve the health problems of Indonesia—those seemed too big to tackle with any resources available to a family or small group. Some satisfaction was taken, however, in the extensive national malaria control campaigns and UNICEF distributions of milk until those programs were interrupted by termination of aid and Indonesia's withdrawal from the United Nations.

All twenty-seven Kenwives reported being generally careful about washing hands and using boiled water. They trained their servants and supervised them on these points. "I was more careful after I had a bad case of amoebic dysentery," said one. A few avoided restaurants; most Kenwives did not avoid eating in Indonesia homes and accepted all invitations, but a few refused. "We ate out a lot, took a 'pill and a prayer.'" With one or two exceptions everyone travelled when possible, but a few avoided travel that required eating in restaurants. Most Kenteam wives had brought a stock of medicines with them but found that most of those they brought were not needed. All but three of the wives consulted the doctor at the American embassy in Djakarta, who prescribed medicines readily available at the dispensary there. (There were occasional periods, notably in 1957–1958, when embassy doctors interpreted the terms of their assignment to mean that they could not properly serve Kenteam personnel because of their nongovernmental status.) Other doctors consulted were the IPB physician, a German doctor in Bogor, and missionary doctors in Bandung.

Language

Bahasa Indonesia—"the language of Indonesia"—is basically Malayan, was widely known in Indonesia before independence, and was made the official language of the independence movement even during the colonial period. However, many Indonesians themselves are still in the process of learning the national language. It is generally considered to be one of the easiest languages for westerners to learn. There are no problems of calligraphy, basic grammar is not unusual, word order is not difficult to control, and much flexibility in usage is permissible. On the one hand, Indonesians are pleased to have foreigners take an interest in studying the language; on the other hand, they are sensitive to its misuse. Many Dutchmen are remembered to have used Indonesian badly while at the same time preferring that Indonesians not use Dutch! During Kenteam's presence in Bogor, many Indonesians sought to learn English, which is officially their second national language, and they preferred speaking English with Americans. When international relationships deteriorated, however, there was some decline in the popularity of the study of English.

Nearly all the Kenwives reported feeling frustrated at first by their inability to understand and use the Indonesian language, and more than half were still frustrated at the time of their departure. As time passed, however, there was some accommodation to the problem; "make-do" skills in communication came into use and the intensity of frustration declined. Ten reported strong frustration after six months, but only three by the time of their departure. All had made some effort to reduce their difficulty; twenty-six had had some instruction in the language after arrival in Indonesia, but only three had before arrival. Adventures in learning the language included talking with servants, bargaining with door-to-door peddlars, talking with sellers in the marketplace, attending classes, using language tapes, vocabulary books, and dictionaries, listening to radio, trying to translate newspapers, using self-study books. "I learned by the sight method and it was often extremely humorous; I would mix up words that sounded similar, for example, ubi, ibu, katjong, kutjing, etc. Fortunately we never had baked mother or fried cat for dinner!" In seeking teachers of Indonesian, Kenwives turned to IPB students and staff, teachers, IPB staff wives, government officials, and others. Thirteen Kenwives felt they achieved quite a lot of speaking skill (only one became very skillful) in Indonesian. Only two thought they had quite a lot of reading skill, and none claimed more than a little writing skill. A majority felt they had achieved at least a little skill in speaking, reading, and writing.

Language was a matter of concern not only to Kenwives but to Kenteam members in general, and their reactions may be considered now in conjunction with the data from Kenwives. Kenteam members had nearly all (thirty-six out of thirty-eight) made some effort to study Indonesian. However, only nine had taken lessons in the U.S. before departure; thirty said they had several lessons while in Bogor; twenty-eight reported self-study. The proficiency achieved was limited, however. Eight reported speaking Indonesian fairly well or fluently, twenty-seven haltingly, and others not at all. Five said they were able to read an Indonesian newspaper. A majority of Kenteam said that language was a handicap, but only a minority of IPB colleagues agreed.

The replies to questioning on this point indicated considerable interest in the matter and presented several rather fully-stated comments from Kenteam members:

Of the thirty-three I knew, only four were unhandicapped by lack of Indonesian. Two of them had prior training with other organizations. It is a shame that AID spends between $40 and $50 thousand annually to keep a staff member overseas and will not spend an extra few thousand to give him the language that would increase his chances of success by at least 40 percent.

In the early years the IPB staff regularly relied on Dutch for their staff meetings and general administrative conversations. Dutch was also the common language in the Ministry of Education. As younger staff members advanced, as participants returned from the States and as students prepared to go to the States, the Indonesians appeared willing and even eager to use English.

Ability to use the language was not essential, but if the team members could use the language it added to the respect of the Indonesians and made daily living easier—for instance in speaking to servants, buying, travelling, and in achieving a better understanding of the Indonesian people.

The difficulty here arose mostly because the students were not able to approach the professors in English. Feeling a lack of self-confidence (basic Indonesian bashfulness) they kept still if they could not express themselves well. So students were not able to approach teachers.

Many Kenteam members tried to take part in semiofficial activities, to encourage personal contacts, to have people in their homes, and to attend student programs and thus overcome the embarrassment and diffidence of the students.

There was feeling among Indonesians that the Americans should learn the Indonesian language, just as they had to learn English in order to study in the U.S. In a sense all Kenteam members were handicapped by their language inability. They could not enter into the full life of their neighbors and IPB. But they "were between the devil and the deep blue sea." For an appointment of only two years, one could not learn the language well enough to be fluent in it. The obligation was more on the host group to improve its English proficiency than on Kenteam members to learn Indonesian fluently. Furthermore, half-knowledge of a language is dangerous; misunderstandings growing out of clumsy efforts to speak a foreign tongue may have worse effects than just sticking to one's own language.

Kenteam members and their wives all made some effort, but the problem of language was not completely resolved and could not have been without more intensive and persistent study, resulting in enough competence for easier communication in Indonesia. It is likely, however, that if all Kenteam members and their wives, or at least a few among them, had been skillful in the use of Indonesian, language frustration would have disappeared completely and no doubt several other frustrations would have been eased, as well.

Revolutionary Atmosphere. In a nation that is newly independent from colonial power the atmosphere is one of revolution. Stress is put on efforts to implement freedom and to build the nation. Ideologies are asserted, acronymic slogans are coined and shouted, the enemies of freedom are named and misnamed, and esprit de corps is fostered by campaigns and attacks. When Kenteam went to Bogor in 1957 a state of war and siege was officially in effect. Travel restrictions and curfews were imposed many times. The stage was beset with symbols and sounds of revolution at an increasing tempo during Kenteam's tenure. Yet, there were twelve other areas of experience in which Kenwives were more frustrated than this! The number who expressed anxiety at the revolutionary atmosphere and slogans in Indonesia increased slightly rather than declined as most of the other frustrations did with the passage of time. Thirteen Kenwives at arrival and seventeen at departure confessed this frustration. This was clearly a response to the acceleration of government-sponsored hostility toward the West, or Nekolim, in the later years of Kenteam's presence in Bogor.

Only four wives expressed fear of anti-American action. Others had various reactions. "A little, only a few times." "No—I did have a feeling of sadness, also indignation, when I found the city of Bandung downtown area blanketed with 'American go home' signs on one visit." "I always felt hopeful. Americans were liked and slogans, anti-signs, etc., were superficial, not genuine feelings." "Several times the chief of party asked us to stay confined to our homes for fear of demonstrations against us."

The comments that Kenwives wrote on their questionnaires explain why the Kenteam frustration threshold was higher with reference to this factor than to others.

I always had hopes things would improve. I felt there was little I could do about it except be friendly with the people.

Since I could not change the tide of local politics, I tried not to worry about it and never discussed it with Indonesians.

It was like watching history on the screen—to see enacted before one's eyes the Cold War, the propaganda, the hate and fear, the slander, the characteristic phrases and pictures, the noise. It was something of a letdown to learn how few communists had dominated so many others until nearly complete paralysis had set in.

I guess I admired the revolutionary ideals. I felt the anti-American slogans were for propaganda, not a genuine feeling in the people.

In our own history, we had revolution. In a "new nation" it is to be expected that there would be strong feelings. My greatest concern was some apprehension over the power of communism that existed then.

It is interesting to note that in listing the categories of frustrating experience, insecurity generated by a revolutionary atmosphere was singled out for emphasis, but insecurity from other causes was hardly more than mentioned. Several Kenteam wives reported problems involving guards (police, watchmen, military personnel), thefts, accidents, fear of unfriendliness, and self-doubts. The comments on guarding did not actually reveal a consciousness of problems, but merely acknowledged the routine presence of the night watchmen, police, and soldiers, with "special guards on special occasions during the revolution and civil war."

The reported incidence of theft was rather high. Nineteen of the wives cited examples, involving mainly small sums of money or "gadgets" of resale value such as cameras, typewriters, clocks, tape recorders, bicycles, hub caps, clothing.

The tone of their comments is merely reportorial and quite undramatic and reflects no strong feeling of fear or insecurity.

Standards of Living, Especially Housing

"Standard of living" is an umbrella concept in that it covers so much, overlapping with poverty, health, and other areas. As identified by Kenwives, it was concerned mainly with housing. Shortcomings in their own housing were sources of some direct inconvenience but even more disturbing were the deficiencies in housing of Indonesian acquaintances. A severe housing shortage existed for IPB staff and many were uncomfortably overcrowded in small dwelling units. Utilities and facilities were deficient. The families of some of the servants lived on earthen floors, without electricity, with recourse only to kali (canal or ditch) water, and with leaky roofs of thatch or tile and walls of woven bamboo-slices or reeds. There were, of course, also some houses of superior quality—permanent structures of cement, teakwood, tile, and other substantial material—but they were also underprovided with utilities and facilities. In relative terms the housing of Americans was superior to that of most Indonesians.

Most Kenwives reported feeling disturbed and frustrated about the standards of living of "the common people." The only change in this frustration between time of arrival and time of departure was that all who felt very strongly at first shifted to the somewhat less intense, "rather strong" reaction during the first six months. A majority reported that they had made some effort to overcome their feelings, the most frequent type of effort being help to servants in solving housing problems. None accepted the situation as inevitable. Twenty-two reported buying, building, or improving housing for one or more servants, supplanting thatch with tile roofs, installing windows, arranging the use of roof water, raising the kitchen work area, providing soak-pit drainage, cementing floors.

Critics of Americans overseas generally refer to the comparative superiority of their housing. Kenwives judged their houses in overall satisfactoriness and only five of the twenty-four said their Bogor housing was comparatively better than their previous housing; five said their Bogor housing was not as good, and fourteen rated it as about the same. For more than half of the wives their Bogor houses were smaller, less comfortable, less attractive, and less convenient. The one characteristic on which a plurality of the wives gave Bogor a superior rating was ease of keeping, although nearly as many said their Bogor houses were harder to keep than their previous houses. No doubt a factor involved in ease of keeping was the help of servants.

Housing inconveniences were specified to include electric power shortages, water shortages, lack of fuel (kerosene), and unsatisfactory stoves. Write-in comments mentioned bathroom deficiencies, need for additional bedrooms and storage space, and too-easy admission of noise and cool winds.

Electric service presented two problems, limited supply and interruptions in service either by scheduled rationing of output from hydroelectric sources during dry seasons, or by storms, overloads, and short circuits. Negatively, this occasioned inconvenience, some food spoilage, and some wear and tear on appliances through fluctuations in voltage. However, gasoline compression-lamps could be used and candles were always available and enjoyed. It was all "part of living there" and "we had much more electric current than the Indonesian faculty." The special problems of electric service were little more than an occasional minor irritation.

There was gas at some but not all of the Kenteam houses. Where available at all, pressure was generally inadequate from early morning to midafternoon. Kerosene was the chief fuel in most of the houses but was often not available in the market. Kenteam members stocked supplies ahead when possible, borrowed from each other, and sent their servants on special kerosene-getting missions. Boiling the drinking water was the minimum operation requiring fuel, and its absence was sometimes correlated with shortness of Kenteam tempers.

Pressure and purity of water supply were also problems for Kenteam houses. Several Kenteam wives recalled irritating periods of inadequate pressure but there was no real deprivation. There were pressure tanks at some of the houses and they were filled at night. For drinking, all water was boiled.

The Bogor municipality had a chlorinating plant, but Kenteam families were advised to boil even the treated water. This was customary among several Indonesian families in the community also, and Kenteam families noted the practice spreading among their servants. A typical comment on water supply was "adequate once you learned when to shower." During seasons of shortage, Kenteam families filled various vessels with water at night when pressure was up and used the supply during the day. Bathing in hot water was practiced by only a few of the American families; the Indonesian custom of bathing two or more times per day with water at natural temperature was fairly common among Kenteam families.

The inferences to be drawn from these replies is that Kenteam wives did not feel they gained or lost much in relative quality of housing in Bogor, given the differences in climate and the availability of servants.

Three-fourths of the wives said they did not feel handicapped in their relations with Indonesians by the fact that they had better housing and a higher material standard of living; one-fourth confessed that they did feel so handicapped. Many supplemented their yes or no response with comments expressing their awareness of a possible problem and their accommodation to it. Reactions to such a question as this, of course, are a function of the personality structure as much as of the external, objective situation, but no analysis of personality can be made here. The explanations of those who did not feel the handicap were rather completely expressed as follows:

The only problem was a constant heartache on our part at not being able to help them more materially.

My Indonesian friends expressed much interest in some of the improvements in our living and in many cases tried to introduce them in their homes. When we first went there, they were astonished at our using things from the pasar (market) to decorate our homes. We explained that not cost, but line, color, and beauty were to be considered. They said they could only use imported or very expensive articles without losing face. Before we left they were using many of their own lovely things. That happened too, in clothes. When we first arrived they always used western materials for their western dresses but as they saw how we used batiks for that purpose, they did too. We tried to keep our house furnishings to a bare necessity and so did not feel that they embarrassed our Indonesian guests.

No, they were used to the difference in American-European housing. For any one whom I thought might resent it, I always said, "Of course, this is your house, really. We are enjoying using it. Thank you, your government made it well."

Some of our Indonesian friends were housed equally comfortably. Most seemed to assume that we were living as closely as possible to the standard to which we were accustomed in the States. Actually, I didn't worry too much about that aspect since I felt that our sincere interest in the Indonesians and our desire to be friends and be helpful wherever we could were most important.

The comments of Kenteam wives who acknowledged feeling a problem were equally thoughtful:

Yes, I was always conscious of this, but overcame my Indonesian friends' consciousness of it by calling on them and accepting with genuine pleasure their hospitality and finding interesting things in their homes to discuss and appreciate. It is human nature to be a bit malu (embarrassed) to invite people who have so much more to share your simple life, but it can be overcome with effort.

Some of our Indonesian friends told us that they felt they could not invite us to their homes because their homes were not as nice as ours.

We always found it difficult to really work on their level because our income was so much larger than theirs. I tried to be very careful with displaying any items purchased which were costly by their standards.

Fewer than half the members of Kenteam (41%) and a still smaller percentage of IPB staff (23%) considered that the houses, cars, and standards of living of Kenteam families were barriers to closer association with IPB staff members. Approximately the same opinion was reported when the reference was changed to "other Indonesian families in Bogor."

However, although they were fewer than a majority, several Kenteam members and several IPB members *did* see standard-of-living barriers. It must be concluded that in some instances, the houses and cars were barriers, in other instances, not. Both Kenteam and IPB members seemed to wish to make their views clear; most Kenteam comments argued that other variables were more important than houses and cars as such. "The fundamental thing was the attitude and the way in which the relationships were set up with Indonesians rather than in the physical fact of houses and cars." The explanations of IPB staff members also attribute barriers to other factors. "The barrier was not determined by the way of living. It was promoted by the amount of mutual understanding. Barrier is caused by lack of give and take."

Explanation of the effect of living standards on association with other Bogor families added little, if anything, to the comments above on association with IPB families, except for a hint of criticism of Kenteam families for the "company they kept" among Bogor people not associated with IPB.

Other Americans Overseas. Americans abroad are highly conscious of each other and are more critical in mutual judgments in places where they are a conspicuous minority than at home. Extremes in behavior are noted with commendation if deserving and condemnation if not deserving! Americans who might not know or see each other in the United States become associates and neighbors at a foreign post. So they become to each other sources of friendship and satisfaction, or they may be sources of irritation, embarrassment, and frustration if they seem other than exemplary in their conduct. What might be overlooked at home, whether virtuous or a bit deplorable, is not likely to go unnoticed in a foreign country. Other Americans can be a frustration, especially to first-timers who haven't yet accommodated to the special values in existence abroad.

A review of matters concerning Kenteam from the beginning suggests that in a group of American families of this size, working under these conditions, a constantly fluctuating combination of conflict and cooperation exists. There are always

223

matters at issue between groups of two or three or perhaps within the whole team. The mobilization of feelings around these issues ranges from mild differences of opinion through petty and temporary quarrels to more deep-seated antagonisms accompanied by anger and lasting longer. At the same time and coexistent among the same groups are common interests and activities, also running the range from incidental and unconscious joint action to reflective and deliberate cooperation, carried out either because of "enlightened self-interest" or because of different degrees of "altruistic" or affectional relationship. In such a group, then, good morale means an absence of tension, quarrel, or even conflict.

In the first weeks after arrival abroad, Kenteam reactions to fellow Americans were apparently less pronounced than responses in some other areas of experience with more immediate impact. After six months, however, apprehensions and a level of concern about the overseas behavior of Americans appeared and tended to endure. Among Kenwives, only eleven were apprehensive at first but after six months nineteen were and at time of departure eighteen still confessed feeling somewhat disturbed. In fact, just half of the Kenwives rated their feeling as rather strong or very strong at end of tour.

Cleanliness and Sanitation. To keep clean requires knowledge, equipment, and effort that have become general and automatic in middle-class American culture. Water may be drunk from the tap because it has been treated; houses and clothing are kept clean by household equipment and practices used by housewives who know about germs. The laws and codes of cleanliness and sanitation are on the books and are enforced by child training and in communities by officers and courts. Communities control water supply, manage disposal of waste, and conduct campaigns for cleanliness and health. None of these practices seem necessary to people in societies where fear of bacteria and knowledge of germ-controls are lacking. Surely Indonesians bathe more often than Americans in general by sousing at least twice daily with water at air temperature, and many an Indonesian has concluded that Americans are dirty. "Is it true that you bathe only on Saturday nights?" But the sousing of dishes in cold water, perhaps at the kali (ditch or canal), which receives also some of the community's untreated excremental output, can be tolerated only by one who doesn't know about invisible filth and the transmission of disease. So the Americans, over-confident in their own traditions of sanitation, and the Indonesians, unaware of the hazards in their practices, disagreed in judgments of cleanliness and sanitation, and this was an area in which some Americans experienced frustration.

Initial anxiety and frustration were common in this area: the amount of frustration declined only slightly after one year, but relaxed somewhat with the passage of time. Eighteen Kenwives felt concern initially, fifteen of them strongly. Fourteen still felt anxious after one year.

Working with Servants. Women in the American middle classes are largely inexperienced in the use of domestic help on any basis other than a few hours weekly for cleaning or laundry. Typically they have done all their own house and family work. To have a houseboy instead of a vacuum cleaner, a washwoman instead of a washing machine, a cook, a yard-man, a nightwatchman, a driver—these are unfamiliar re-sources and many a newcomer among Americans abroad has toyed with the idea of dismissing all servants to regain privacy and personal control over the menage. The standards of native servants, trained by the Dutch or by families in the Indonesian elite or not trained at all, are very different from those of an inner-directed American housewife with respect to cleanliness, time-schedule, meal service, and internal ar-rangement of furnishings.

Twelve Kenwives had employed no servants in their Amer-ican homes, and ten had employed only one, part-time. Four of the wives had employed from one to five servants during previous residence overseas. In Bogor, none of them had fewer than three servants; seven kept three (houseboy, cook, and washwoman), ten kept four, seven kept five, and two kept six servants.

Working with servants was a matter of anxiety and concern to half of the Kenwives (fourteen) and ten of the group expressed strong anxiety. The intensity of frustration soon declined (at six months and thereafter) but the number of women affected increased to twenty by the end of the first year and was down again to twelve (nearly half) at the time of departure from Bogor.

The collective attitude toward having servants was ambivalence. When asked what they liked about it Kenwives mentioned the ease of entertaining, freedom from the routine of preparing meals and keeping house, and the opportunity of meeting village people through servant families. That servants could use native equipment, do the washing and ironing, and keep the house clean at all times was also enjoyed. Being relieved of buying in the market, having a driver, having a cook, having someone to wash dishes and someone to be with the children—Kenwives also mentioned these.

I was able to be a relatively effective human being rather than a robot managed by the strings of family needs.

It made me feel rich for a while—gave me time for something besides cleaning, cooking, etc.

It made me realize that I preferred teaching and study to staying home and taking care of a house.

It gave me a much needed two years' rest and broadened our knowledge of "how the other part of the world" lives.

I feel that I got to know the Indonesian people better because of my close relationship with my servants. It gave me the opportunity to learn much more of their thinking and customs and my respect for them grew.

We had the same servants for three years and considered them as friends more than servants. All three took pride in doing things well, and seemed eager to do extra things to please us when they knew we liked them. There were no problems in any real sense.

They helped our children learn to respect everyone as human beings just as they were.

But there were aspects of having servants that the American women did not like. Lack of privacy and having to be responsible for the problems of servants and their families were mentioned most often. Other aspects listed were the need for constant supervision, "having all those people in the house," being unable to trust them, a preference for doing one's own work, "watching the cook" about cleanliness and water boiling, trying to teach cleanliness, their low status, and the bad influence on children (making them "lazy"). Theft was occasionally a problem. One Kenwife said, "It was most discouraging to be 'taken in' after so much help had been given. These people live so simply that our homes are like palaces for them. I think they carry over the village philosophy of sharing with all and they feel we won't miss the things or can buy more and they have so little. Still, when it happens my 'Yankee code' is offended."

Most Kenwives took personal responsibility for servant families, thinking of their needs for medicine, clothing, food, shelter, dental care, divorces, sickness, weddings, funerals, new babies, eye care. "Their problems became my problems." All but three provided rice in addition to a wage for their servants; all paid for doctor fees and medicine, allowed servants to eat at the Kenteam home, and gave money additionally. Only eleven provided milk for the servant's children. All but three made loans to the servants. These were for buying, rebuilding, or remodeling houses, for bicycles, doctor bills, funerals, weddings, gold earrings (savings), travel, school books, selametans (ceremonials), clothing. There was a thin line of distinction between loans and gifts; money given to servants was for the same purposes as the loans and for additional items such as divorces, festivals, births, and driving lessons. Other gifts included utensils, clothing, eggs, radios, wrist watches.

When asked whether having servants had any noticeable effects on their children, thirteen of the Kenteam wives said no. The others mentioned a few types of effects, three of them positive: the children were well taken care of, they liked having someone to play with, and they learned the Indonesian and Sundanese languages, as well as native games and dances. Negative comments included "Made the children over-bearing, bossy," "children not as cooperative in caring for their personal things as they were at home," "made them somewhat lazy," and "spoiled" them.

Half of the Kenteam wives said their Stateside return to a
"no-servant" pattern of life did or would require some adjustment but that this would be only a temporary and minor
problem. "It was difficult to do everything for yourself for
a month or two." "Yes, I've become lazy with respect to
house cleaning." "I'd gotten soft, and tired more easily and
took longer to do tasks than in pre-servant days." Most Kenteam wives maintained contact with their servants after departure from Bogor by sending money, letters, family photo-
graphs, and other items.

The reaction of Kenteam wives to having servants was not
couched in judgments of the social structure in which there
are positions for domestic service, but was an analysis of the
role of servants in the immediate experience of the American
family. The comments reflect a general appreciation of the
novelty of the experience, a feeling that they would have been
unable to cope with circumstances of life in Bogor without
them, and a general gratitude for what servants contributed
toward their experience, without overtones of criticism of
either the servants or the system which provided them.

Time Values. In American culture it is easier than in Indonesia to control behavior in relation to time, and the value
of time is expressed in the saying "Time is money." A characteristic reference in Indonesia, on the other hand, is to
"rubber time" (djam karet) in good-humored description of
the irregularity and unpredictability of time expectations. To
keep a promised appointment anywhere requires control of
communication and transportation. If these do not exist, dates
may be made but skipped. Invitations may be accepted but
always with an implied recognition that attendance may not
follow, that plans may be changed without prior notification
or subsequent explanation. In a culture which has made this
accommodation, speed has no intrinsic merit. How can speed
be important if it is impossible? And how can any attitude
other than patience be appropriate when the passage of time
must be accepted and cannot conceivably be altered? In fact,
the uses of time and other resources seem beyond control.
"Sufficient unto the day is the evil thereof." The spirit world
or the dukun (shaman) may be asked to prophesy, and it is

acceptable to invoke his concern for a time to come. But the orientation to the future that would produce savings, insurance, investment, or long-time planning can come only in a culture that has increased its controls over time and its confidence in their use.

Americans value punctuality, regularity, control, predictability, acceleration, saving (or efficient use), and looking ahead. Indonesians traditionally do not have these values; they regard patience very highly, they approve deliberateness and moderation in action, and they accept flexibility. They do not require regularity or predictability; they do not stress economy of time or anticipation of the future. So the time-saver, the clock-watcher, the man-in-a-hurry, the time-budgeter is likely to face devastating frustration in Indonesia or in any of the less-developed societies, where time has been accepted as given or as allocated only by fate. In the advance of economic development, time becomes a scarce resource, and time-saving innovations are a sign that modernization is under way.

Most Kenteam wives confessed irritation at the cultural differences in the meaning and use of time. Only four (about one in six) were undisturbed about time on arrival in Bogor, and half were strongly frustrated. These feelings declined in strength during the first years of residence, but no further change occurred thereafter and time frustration never disappeared completely. In orientation to the social meaning and use of time, Kenwives adjusted, but they were surely not completely assimilated. Conversely they reported several indications of the acceptance by Indonesians of American time values. An effort to acquire patience was one of the adjustment procedures reported by Kenwives.

I tried to find out how the Indonesians did think and to fit my plan to theirs. At first there were a couple of instances where the number of people coming to dinner did not jibe with the number invited, sometimes more, sometimes less, but I tried to adjust my thinking to it. I felt as we stayed there that the Indonesians were adjusting to our ways in regard to their engagements.

I talked about it with friends in my classes who became more punctual. I set schedules for dinner parties up one-half hour to get people there closer to on-time. I planned flexible time myself.

I found my degree of feeling didn't change but my attitude toward the other fellow's position on time became one of understanding about lack of transportation, newness of the time concept (from the agrarian following the sun), the lack of communication, the great neighborliness which imposes on time. High etiquette standards require finesse, which takes time. To be Indonesian proper is more important than promptness.

I loved the peace it gave me. My husband says I never got over this. The Indonesians taught me patience and tranquility.

A majority of the Kenwives made an effort to demonstrate or teach punctuality, promptness, or time-scheduling. They did this by setting an example: deliberately being on time. They instructed their servants and they exhorted prospective participants who were preparing for study in the U.S.

The tropical nature of the time schedule (official hours at IPB were 7 A.M. to 2 P.M.) accounted for the fact that Kenteam members were in their homes and with their families more than would usually be the case in the U.S. Nineteen of the wives reported this to be the case.

Most Kenteam wives denied that they and their husbands were any more tense overseas than at home. They said they were more relaxed; there was less pressure of time, housework, "keeping up with the Joneses"; more free time, more time together. "As a family we found it to be a wonderful, warm, and unifying experience and learned much about enjoying every day for itself. This we learned among other things from our Indonesian friends." There were eight Kenwives, however, who reported *more* tenseness overseas, but due not so much to pressure and frustrations involving time as to anxieties over other matters. "We felt that all we did was observed by our Indonesian associates. We were foreigners in a strange culture so it was inevitable that we would feel more 'tense' than at home. We were eager to make a favorable impression, but of course we made mistakes despite our efforts."

Etiquette in Daily Life. Highly formalized systems of ritual have been practices in the various cultures and ethnic groups of Indonesia, and there are forms of language and gesture to be followed strictly at given places and times. Nationalism brings some strain of secularization and uniformity among the cultures, and foreigners are relatively exempt from requirements to be meticulous in the niceties of tradition. However, the observant American newcomer is soon aware of conspicuous differences in custom. In place of the handshake of greeting, there is the sembah gesture in which one's hands, placed palm to palm, move during the bow of acknowledgment from one's face to meet—or almost meet—the fingertips of the other, who is performing the same gesture. There is the ever-present smile, required even in anger or grief! There is the voice, so muted in respect as to be inaudible at first. There is the proffered drink which a guest does not consume on receipt but only later at the host's special invitation. There is the indirect statement, couched in humility and respect and conveying only messages that it is hoped will please the auditor! There is the avoidance of any situation in which one might have to say no. There is the restricted use of the left hand which some functions make unclean for other uses.

The Indonesians with whom Kenteam worked, of course, were more familiar with American etiquette than Americans were at first with Indonesian etiquette. Situations were frequent in which the Indonesian made his approach by courteous use of American gestures, while the American responded with an attempt in Indonesian forms. The embarrassment of uncertainty was often present on both sides, however, and reluctance to face embarrassment surely influenced much interaction between Indonesians and Americans. Generally, the adoption of Indonesian practice by Americans was somewhat casual and of limited depth, but the effort of Indonesians often led them farther into following American practice than occasional courtesy would require.

Nonfamiliarity with etiquette was initially a matter of frustration for about half (thirteen) of the Kenwives. The number declined somewhat during the first year, but one-fourth (seven) reported still having this reaction at end-of-tour. In strength, however, etiquette-frustration was of minor importance even initially, and most Kenwives were rid of it quickly as a strong irritation. After a year, only one, and at time of departure, none at all, reported concern.

The ladies of an Indonesian-American group "were most

gracious in sharing with us simple customs of the country." This association was formed in the first year of Kenteam's presence in Bogor, taking on the nature of a homemakers' group in the United States, although it could never organize officially as a club. In the earlier years a lively mutual interest in sharing customs was evident and an effective orientation was thus available for Kenwives. In later years, when some internal Kenteam traditions had formed and newly arriving Kenwives could be initiated by those already in Bogor, there was some decline in interest. The newcomer Kenwives were disposed to criticize the self-selecting, restricted Indonesian composition of the organization and seemed at times insensitive to the threats an Indonesian woman might brave to associate closely with Americans. Regulations against formal organization actually prevented the overt continuation of this club, though informal contacts continued throughout the period of the affiliation. In each cohort of Kenteam wives there were at least a few who maintained association with the original Indonesian sponsors of the relationship.

Population. It has long been known that Java is among the most densely populated areas of the world, and that recent population increases have exceeded those of former years. Efforts to promote migration from Java to other islands have been continuous for half a century, but they have not persuaded any large numbers of Javanese to go so far that they couldn't return annually to visit family graves. Facilities do not exist that could remove from Java the number of persons by which its population increases. Agricultural involution (absorbing more people into the same farming system) and movement to cities have not alleviated the pressures of population in villages. The undiminished rate of increase and the absence of social concern with the problem are sufficient to alarm American observers. Interest in birth control and family planning was starting slowly and surreptitiously, not coming into the open until after Kenteam's departure, and full acknowledgment by the new regime in national government of the seriousness of the problem.

"Anxious concern for population problems in Indonesia"

was reported strongly by nearly half (twelve) of the Kenwives at first, but the proportion dropped somewhat as time passed. Only three said they had taken any action to overcome this alarm and apprehension. Review of comments made, however, reveals that several Kenwives did take various kinds of action.

I provided reading material and studies and provided contraceptives if requested.

We only gave advice and helped friends if they asked us to. I had some books on birth control written for use in rural villages in India which I gave friends when they asked about birth control. We also gave contraceptives to those who asked.

While I saw the problem, was actually aware of it, I again felt that it was not my business. If their religion and the president recommended large families, that is their concern, not mine.

What could I do? I made a contribution to a world organization and asked for information on what was available locally, but got no reply. My "pills" would have cost them more than their monthly salary.

Cultural Values. Differences in values between cultures, if not in critical areas of experience, add interest and zest to cultural interaction; in critical areas they may bring frustration and conflict. Americans, for example, value individualism and competition; Indonesians value mutual assistance and cooperation. Americans value decision; Indonesians value consensus. Americans may sacrifice beauty to efficiency; Indonesians may sacrifice efficiency to beauty. Indirect communication by Indonesian values is courteous and proper; by American values it may be deceitful. Indonesians value importance and status; Americans value achievement. Self-effacement to an Indonesian conveys respect; to an American it may be interpreted as lack of initiative. So, the impact of foreign values on American overseas wives is potentially disturbing or potentially exciting, and Indonesian-American contrasts in values were important to Kenwives. Just over half of them (sixteen) reported disturbance and frustration over important features of the Indonesian value system, and the number who were thus disturbed did not change importantly. As time passed, only six reported any specific action to overcome this problem but several described accommodations which occurred. Some studied Indonesian values and beliefs in an effort to understand what was initially strange to them. For a few, study resulted in increasing dislike; for others it brought growing acceptance or at least tolerance.

Money. In money, any middle-class American is rich among the people of comparable position and status in a "new nation." The command of dollars over local currency is without challenge or parallel. Legal rates of exchange are favorable to the buyer with dollars and illegal rates are sometimes fabulous. The American's income in local currency is many times that of his Indonesian counterpart. He can buy without hesitation what his friends may never aspire to own. He can travel, entertain, shop, equip his home, dress his family, enjoy books, magazines, records, educate his children, buy expensive foods —all without much change in his usual volume of consumption, except that what he buys abroad is probably cheaper per unit that what he buys at home and of somewhat exotic character by virtue of its foreign origin and other-culture reference. The position of relative affluence is new to the American technical man abroad, and the oddness of it all may engender various kinds of frustration, annoyance, or irritation.

Money was a source of worry at first to fifteen Kenwives and still troublesome at time of departure to eleven; there was a slight decline in the number strongly disturbed, but the number reporting weak feelings of concern remained about the same throughout the period of residence in Bogor.

Almost all the women (twenty-five) said they learned to bargain (tawar), declining high asking-prices and offering very low payment, later agreeing to a compromise sale-price. Bargaining etiquette in Indonesia requires approval by the seller; without his consent to dicker, the price must be presumed to be fixed. He may offer the privilege of bargaining or it may be requested, but traditionally the process is not sanctioned unless the seller consents. The motives and skill of Kenteam wives varied. "Sometimes I let the tukang win and sometimes I would win. If he did he usually gave me something extra for the pleasure it gave him to beat me. I bargained more to be consistent with Indonesian custom than for lower prices."

Kenteam wives in general (but with exceptions) paid wages and prices more generous than the prevailing rates. Decisions in this matter were highly rationalized, and called forth several types of justificatory explanations. All the Kenwives reported buying more consumption goods, art objects, and so forth, than they would have in the U.S., and gave an interesting variety of explanations. Objects bought occasionally or often, in order of popularity, were batiks (dyed fabrics), woodcarvings, paintings, jewelry, Djogjakarta silver, Chinese porcelain, Bangka tin, wayang pieces (shadow play or puppets), leatherware, and antique furniture.

But comments on bargaining, paying high wages, and buying consumption goods, while showing accommodation of Kenwives to features of the Bogor market, did not reveal their frustrations. These come out more clearly in other remarks.

Money rates—legal, blackmarket, "accommodation"—were very confusing and unrealistic. They also created "gulfs" among our own people of different status.

Inflation bothered me most in the effect it had on the Indonesian who had no favorable exchange rate.

Money was a constant irritation—feeling cheated and cheater simultaneously. I never became at ease or resigned to the situation.

Food. The foods and food habits of most Indonesians differ from those of most Americans in several features, but perhaps the two most evident to a newcomer are the ubiquitous dependence on rice and the use of hot seasoning. The meat-and-potatoes-and-bread pattern of rural and middle-class Americans, with the balanced additions of vegetables and fruits, can be simulated, even reproduced in large measure. Nearly all the ingredients were available, in some variety or other, in the Bogor markets, supplemented by the treats bought in the American commissary at Djakarta. Some special knowledge of the intricacies of the pasar and the skills of bargaining were required, however. But the "rice table" menu or Chinese foods are the characteristic fare, often hotter in taste and cooler in temperature, than the usual fare in an American home, and Indonesian cooks will prepare them unless otherwise instructed and supervised. Among Americans, tolerance of what they find unusual in food varies greatly; some families more than others would find Asian meals and diets to their liking.

Most Kenwives (sixteen) reported being anxious and frustrated by food problems on arrival at Bogor, but most of that anxiety disappeared before departure (eighteen had no problem then). The concern over food began to lessen soon after arrival; there were as many nonfrustrated as frustrated Kenwives after six months, but by the end of the year none were left with strong anxiety. Everyone had eaten and lived! About a third (eight) of the Kenwives reported efforts to overcome food frustration and only four avoided eating Indonesian foods. All but six of the Kenwives reported that they liked and learned to prepare Indonesian foods.

Nonavailability of some kinds of food was mentioned but it was not considered a disturbing problem by most Kenteam families. Team members in the first years (1957–1960) had to purchase on the local market or import their foods. In the later years (1960 and thereafter) membership in the American embassy commissary was extended to all. Half of the Kenwives responding (twelve), those serving in the first years, could not have used the commissary, but fifteen (those serving later) did make purchases there and enjoyed their contribution to food supplies.

Asked what Indonesian foods they liked best, none of the twenty-six wives failed to list at least one; when asked what they liked least, five wives mentioned none. Thirty preferred Indonesian foods were named, plus fourteen disliked foods. The great favorites were fruits (lumped together), saté (chunks of meat and vegetables on a stick broiled over an open fire), and nasi goreng (fried rice), each volunteered by eleven of the twenty-six wives. Other foods got from one to three votes apiece. The most disliked food was the durian, a fruit which has a prominent odor that serves as a repellant to some, but to others is no barrier to the enjoyment of its unique and pleasant taste. Yet three wives checked it as a favorite food.

Pepper-hot foods were singled out for dislike by some. Twenty Chinese foods were specified as "liked" but only six as "most liked" by any of the Kenteam wives. The best known or most often mentioned were tjap tjay (chop suey) and sweet and sour meat.

The result of this little survey does not add up to a preference for Indonesian or American foods, but the ratio of likes to dislikes is two to one. There were seven Kenteam wives who did not enjoy Indonesian foods, and only one did not enjoy Chinese food.

Food topics that induced anxiety were really only food-related: inability to satisfy the family's food preferences was of minor consequence and was readily overcome. Many new satisfying food tastes were formed. The troubles at home were the problems of sanitary food handling, satisfactory cooking, and the always latent threats to health. The troubles outside of the home lay in the observation of hunger and bad diet vexing the lives of the Indonesian multitudes.

Climate. The climate of the tropics is beneficently frustrating for the newcomer adjusted to the alternation of seasons and rigors of temperate zones. The equator crosses Indonesia and Bogor is only a few degrees south of it. The relative year-long uniformity, the lack of distinctive seasons, the heat of the sun, the monsoon deluge, the mold in the clothes closet, the spoilage of foods—these were all inconveniently but not dangerously different for Americans who could only have read about them before. Paradoxically, what is different may be both attractive and pleasant as well as a source of discomfort and irritation. A little above sea level (700 feet) and close to the mountains (Puntjak), Bogor splits all the differences and presents a remarkably pleasant climate to anyone who stays long enough to learn that even a temperate-zone man may be comfortable there. Very few spots in the world have higher annual rainfall but the weather is clear longer than it is rainy—even on most of the rainiest days.

Equal numbers of Kenwives reported feeling and not feeling frustration because of climate on arrival in Bogor. The number feeling no frustration remained about the same at departure as on arrival. So those who were troubled initially never completely overcame the irritation or discomfort, although most of them reported some effort to adjust. Those who still felt frustrated at the end, however, reported only weak or minor irritation. After the first weeks in Bogor, none were strongly affected.

As a defense against humidity and mold, "hot-closets" (with electric lights always turned on) were used by a few families and dehumidifiers by two or three; about one-fourth of the families used room air-conditioners. Most used no more elaborate mechanical defense than electric fans, which were provided in all the houses. An adjustment widely adopted, however, was the afternoon siesta. Twenty-four of the twenty-eight Kenwives said they took an afternoon rest, either sleeping or remaining quiet. They explained that this was required by the heat. "I had said that I would never 'nap' in afternoon. Boy, that was a big one! Yes, I slowed down."

Conclusion: Concern for Self and for Others

The list of frustrations discussed above was drawn up by the wives, the data were submitted by the wives, but the inferences may safely be extended to the whole Kenteam population. The wives may not have been a random sample, but they were a discerning and coherent one.

The categories are of different order and it is puzzling to classify them, to find their "rhyme and reason." They are a mix of hazards to the person frustrated and of worries about others. Deprivation of news was personally threatening. Poverty was a threat to Indonesia. Frustration over the former was ego-centered; frustration over the latter was other-directed. And herein lies a suggestion to those who would care to refine the concept of cultural shock. Ego is shocked by threats to itself and by threats to others, and this contrast of concerns, this ambivalence of self-concern and concern for others, egoism and altruism, runs in varying proportions through the list of Kenteam frustrations. An important inference is that the threats-to-self-and-family are lower on the list than the threats-to-Indonesians. The objects of frustration are as good as any proof yet found of the sense of mission that was a major propellant of Kenteam.

10. Retrospect

The murkiness of recent Indonesian history and the continuing fragility of political and economic order make it doubtful whether criteria or evidence to determine the effects of the Kenteam project could be precisely identified. However, a worldwide study of technical-assistance institution-building programs, the CIC-AID Rural Development Research project, was started during the last year of Kenteam's work at Bogor, and its findings, made known in 1968, present a generalization about the sequellae of contract termination.[1] By survey of the institutions which had received technical assistance in East Asia, the authors of the report concluded that none of those from which U.S. universities had withdrawn "has achieved the kind of overall maturity" necessary for "self-generative performance"; that the physical criteria (buildings, numbers of faculty and students) did not correctly measure maturity; that each developing institution after termination suffered a "traumatic interlude," retrogression, and loss of competence; that there was need for continued assistance in "key areas."

The report of the CIC-AID survey offers ten general recommendations, each elaborated in specific subproposals, for improved AID-university cooperation in technical assistance. These recommendations have considerable strength because they are drawn from the study of sixty-eight contracts. This is proof, at a certain level, of their validity and importance. However, every one of them could have been proposed from the experience of the Kenteam case alone. The recommendations are, in thumbnail brevity: stronger commitments to expanded and longer programs; more flexibility in contracts and better liaison among the parties; more study-and-application of the institution-building process; use of the basic ideas which were applied in the U.S. to the land-grant system; wider host-guest participation in planning and review; changes in features that "turn off" American university professors and administrators; better orientation of the Americans involved; better participant programming; more university activity to promote public understanding of the far-away need and task; and strengthening the capabilities of American universities.[2]

It is clear that the termination of the contract which took Kenteam to Bogor in 1957 and brought Kenteam home in 1966 occurred under special circumstances and that the affiliation could not have been prolonged, whether needed or not. However, in remembering American aid programs in Indonesia during the Sukarno period, observers generally acknowledge that the university contracts were probably the most successful of all.

Higher education in the agricultural sciences in Indonesia was not the only underdeveloped resource. The concepts and strategies for university involvement in technical assistance were also incompletely formulated. Since the termination of the project at Bogor and since the writing of this book, there has been progress in the testing of outlines for studying the institution-building process, and one frame of reference which has been discussed in developmental assistance circles is available to apply in a retrospective judgment of Kenteam's work.[3] This model presents institution-building as *innovation,* accomplished by *organization* within an *environment,* by establishing certain *linkages,* and directed by *management.* These are five "basic concepts"; they are fulfilled by designated elements and results are then evaluated by selected categories which are part of the model.

The innovation was the Agricultural University—IPB—developed from two initial faculties at Bogor, agriculture and

[1] CIC-AID Rural Development Research Project, *Building Institutions to Serve Agriculture* (Lafayette, Indiana: Purdue University, Committee on Institutional Cooperation, 1968). Because of the deteriorating political situation at the time, however, the Kentucky project at Bogor was not among those visited by the CIC-AID survey staff.

[2] CIC-AID, *Building Institutions,* pp. 4-23.

[3] The framework introduced here is modified from a presentation at Purdue University in August 1969 by Dr. Milton J. Esman, director of the Center for International Studies, Cornell University. Dr. Esman and others have sometimes alluded to this as the Pittsburgh model because of its formulation by groups in which University of Pittsburgh representatives were centrally involved.

veterinary medicine. The organization was the technical-assistance complex formed by the two national governments and the affiliated institutions of higher agricultural education. The environment was what has been described above as the Indonesian "situation." The linkages were relationships set up between IPB and various other agencies and organizations concerned with its work, including especially other colleges in Indonesia, the national research institutes, and the extension division in the Ministry of Agriculture.

Further analysis using the foregoing concepts shows clearly that the purpose at Bogor was innovative: to develop an Indonesian center of higher education in the agricultural sciences on the base of two University of Indonesia faculties. But the purpose was never completely formulated: the image of what was to be innovated was blurred and parties to the project had different goals in mind. For what area would IPB be a center: West Java? All of Java? All of Indonesia? Southeast Asia? What were the time horizons: How far was development expected to go in three years? Five years? Ten years? Indefinitely? Higher education at what level: Doctorandus or baccalaureate? Insinjur or master of science? Doctoral? Post-doctoral? Higher education with what scope: Would research, extension, public service be included? Would fisheries, forestry, and various technologies be included?

A defect of strategic planning and program development was that the IPB goal was never clear enough with respect to specific points. Yet, it would have been a mistake to ignore the inevitable intervention of unpredictable factors and to be overspecific in stating long-time goals. It would have been a mistake also to wait for Indonesia to clear up its national goals before settling on purposes for IPB, although the direction of national development would surely influence the climate of expectations for IPB and the diversion of resources to IPB.

The popular insistence on clarification of goals before operations begin, in order to facilitate evaluation, is proper; one cannot, having mounted a horse, "ride off in all directions." But this insistence can also be a disguise for delay. Sometimes after work is started and progress is thought to be slow, the complaint that goals were not clear enough may be merely diversionary. It should not be expected that all objectives can be determined in advance. Indeed, purposes must be stated, but sometimes when allegedly not "spelled out" they are implicitly present, may only require explication, and, the continuance of work need not be retarded. Part of the developmental process is the preliminary specification of goals; another part is their clarification and revision as new considerations come into view. This is really the way things got started at Bogor but too many of the goals remained implicit and unformulated. No doubt it would have been better if the parties to the original agreement could have anticipated more fully the necessary dimensions of time and scope, leaving less room for uncertainty and less reliance on the improvisation of short-range decisions. Deficiencies in strategic planning left too much for operational planning to "pick up."

The institution-building goal of the project was, indeed, its main specific purpose—its manifest *raison d'être*. There were several associated purposes on the American side—latent or covert in character, not written into the contract—and an appraisal must be made also of the degree of their achievement. One was to further the diplomatic interest of the United States at that time by strengthening the foreign assistance program, which in Indonesia was intended to keep that new nation from slipping over to communism. At a different level, one purpose was to pull the resources of American universities, in this case the University of Kentucky, into the governmentally sponsored international programs for economic development and to deepen their capability to participate effectively in advancing the progress of developing countries, in this case Indonesia.

With reference to the first latent purpose, the whole American assistance program nearly failed at the moment of the attempted coup in the fall of 1965, but the Kentucky contracts were the last elements of the AID program in Indonesia to withdraw as the political clouds darkened. The near-miraculous turnabout of Indonesia's political direction and the shape of her economic planning thereafter brought institutional segments and persons of noncommunist commitment to the top.

Many among them had been students of American development, recipients of American education, and functionaries of the agencies and programs in which American effort had been expressed. There is no doubt whatsoever that Kenteam contributed heavily toward the achievement of the American diplomatic purpose.

With respect to pulling the resources of the University of Kentucky into the work of development abroad, it is clear that the university, as contractor, made a substantial contribution to developmental effort in Indonesia. It was more in the name of the university, however, than by deep involvement. Forty percent of Kenteam's members, for example, were University of Kentucky "regulars" but the rest were part of the university only for the duration of their service under the contract.

The university emerged from the Kenteam experience only a little, if any, more capable than before of technical assistance abroad, and it is a criticism of the contract, rather than of the university or AID, that not more was accomplished toward this purpose. The reliance of state universities mainly on state support makes it difficult for them to develop international expertise and add expensive international programs without special programming and financing. The studies cited above by CIC-AID, committees of AID, and the National Association of State Universities and Land Grant Colleges (NASULGC) have identified the major problems involved, and it need be noted here only that the case of Kenteam confirms the need for building up the developmental capabilities of American universities.

In considering organization, it is appropriate to single out one item for special comment, namely, the team. Kenteam was a dominant feature of the organization-for-innovation in Bogor's institution-building, and reference to the guest Americans as a team focusses attention on their identity as a group, their association with each other, and their common attachment to the project. Connotations of the word team, moreover, stress internal cohesiveness and a boundary which separates the group from other groups. This separate visibility of a team is heightened if there is formal identification of a leader, a "chief of party." The idea of a team is as American as the word, and herein lies a difficulty which Kenteam recognized and sought to overcome—with intermittent moments of success. The basic group in this whole institution-building effort was the personnel of the developing university, including both host and guest members. Hence, the basic membership of a guest professor was in his IPB faculty and his IPB department, rather than in the group of his fellow Americans. But while the former had priority in strategic importance, the latter had greater ease of performance. In perplexity or stress it was natural to turn away from the stranger-host and toward the fellow-American guest.

Foregoing discussions have noted that Kenteam itself, in each phase of its experience at Bogor, sought to minimize its separateness of identification and that the Indonesian hosts were aware of and appreciated this effort. But to the home university, its personnel in Bogor were a team; to AID in Washington and Djakarta they were a team; their work was interpreted as a team effort, and herein appears a problem of some subtlety. Cohesiveness among the Americans was undeniably essential and there were needs also for some degree of inter-American loyalty, shared thinking, mutuality in reinforcement of morale, common commitment, joint planning, and just plain friendly association. These features are indeed recognized in the concept of team.

Beyond this there is some danger in referring to the group as a team. If the group were to become a unit of action (like an athletic, military, or survey team), the character of its work would be inaccurately identified. It was important that Kenteam not be a decision-making unit, acting by pressure of majority vote, resolutions, or "team" recommendations. Although the Pittsburgh model of institution-building was not yet known, its main principle was applied in advance at Bogor —that in a case of innovative technical assistance, the agents of change are the host nationals, not the guest specialists.

The designation of a leader or chief of party complicated the incumbent's problem of identity in relation to all other parties to the institution-building effort. In the complex of relationships involving several agencies, the "team leader's" responsibility seemed ambiguously diffuse. A dean on the home campus might define his role as that of a department head; an AID director in Djakarta might construe his role as that of an officer of the mission; an Indonesian official might consider him the American counterpart of a dean or occasionally a rector. Actually, he was none of these and the position should have had better definition. It was a new position, unique to the task, unlike any other in the organizational chart of a university. The basic role was actually that of *representative of the home university* in all matters relating to performance of the university's mission under the contract. Confusion over responsibility could be cleared up and any question of propriety in communication could be answered by reference to the status and role of a representative of the home university.

The Pittsburgh model moves from the five basic concepts listed above to designated elements of institution-building, the first being *leadership*. In this case the dominant leader was a dean of agriculture who became minister of education. There was a swirl of ideologies around the developing institution, but the dominant doctrine was belief in education as basic to development. The main *programs* were the guest professor strategy, the formulation of curricula and the award of degrees to graduates, the purchase of needed facilities, and the provision of advanced study abroad for Indonesian faculty members. *Resources* included mainly the funds from American foreign assistance and from Indonesian financial allocations, but also the personnel and their expertise. *Internal structure* developed in the networks of departments and colleges that were organized.

Hopefully, the planners of future institution-building programs will apply the foregoing frame of reference, or improved versions of it, in outlining the problems identified, the innovations to be introduced, the organization required, the relevant aspects of environment to be understood, the linkages to be made, the management to be undertaken. Hopefully, they will then proceed to build their plans around these basic concepts with deliberate attention to leadership, doctrine, programs, resources, and internal structure. Then when universities are sponsored to join as host and guest, they can select, orient, and assign personnel to their staffs, and they can check progress and revise procedures within the specific and relevant format of a scheme for institution-building. The prospectus for a project can be written, the contract drafted, the plan of work formulated, reports written, evaluations made, revisions decided—all within the relevant categories by which institution-building is defined.

It remains in this discussion to apply the final set of categories from the modified Pittsburgh outline introduced above—those which indicate whether the end is being achieved, that is, whether an institution has been or is being built. These indicators are *survival, intrinsic value* (measured by *autonomy, influence,* and *recruitment*), *normative spread,* and *innovative thrust.*

What about IPB, in these terms, after nine years of the affiliation? The Agricultural University had been established and was surviving. The acknowledgment of its intrinsic value was manifest by its relative autonomy in developing programs, by its influence in Indonesia's agricultural development, and by its recruitment of funds and of staff. Its normative spread was seen in the replication of its curriculum, rules, and staffing patterns in other Indonesian colleges of agriculture. Its innovative thrust was evident in its development of graduate-degree programs within Indonesia and its affiliation with faculties developing in other provinces. The tentative conclusion can be reached by initial use of these indicators that the end goal of the institution-building project was partially achieved. This is technically—but minimally—correct and a glance at the status of IPB in 1968, and of plans being made then to resume technical assistance, will help to verify or qualify the conclusion.

Visiting Bogor personally and briefly in the early summer of 1968, three veterans of Kenteam service (the author, who was Kenteam's last chief of party, his wife, and a former American secretary of Kenteam) were able to note certain qualities of the situation two and one-half years after the 1966 departure. They were met with a surge of affectionate hospitality that suggested a country-style American family reunion during old home week! The officials of IPB extended official greeting and the former chief of party was ushered to an office and a desk with his name on it, ready for service in his "old department." The faculty held an honorary picnic-luncheon at Darmaga, participants and their wives called with their newest babies, the groups of faculty women who had met with Kenteam wives reassembled for the occasion, and three of them had festive dinners. By all the social courtesies which could have been employed and with all the symbolic directness that their culture permits, the people of IPB told the people of Kenteam that the "new order" is grateful.

The visitors found there had been no easement of the economic need; IPB staff still had to scramble in every possible way to supplement meager wages; everyone sought to "moon-light." There was no money at IPB for repairs or replacements or the routines of current operation. Even the cost of stamps to put on mail to American friends was more than either the personnel or the institution could bear. If there had been only a little more improvement in the economic context of life, to go along with the political turnabout, one might have expected prominently visible evidence of growth and development at IPB.

Even so, it had been encouraging in Djakarta to identify among the members of a national workshop on food,[4] several IPB staff-member scientists who had been Kenteam's colleagues and participants. It was encouraging to determine by count in Bogor that 96 percent of the participants who had returned to Indonesia before Kenteam's departure were professionally employed, 70 percent of them still at IPB. It was gratifying to experience the outpouring of warmth and friendly ap-preciation of IPB colleagues and to feel the depth of greeting from the Bogor community. With a sense of privilege, the hospitality of Indonesian friends was accepted and the com-forts of a house once occupied by Kenteam Americans was enjoyed. There was dramatic climax in a meeting of welcome especially arranged by the rector, his associates, and the deans

[4] Sponsored by the Indonesian Science Institute (Lembaga Ilmu Pengetahuan Indonesia) and the U.S. National Academy of Sciences.

who, in recalling the SAD incident which had once posed the threat of ejection, erased the 1965 censure by an expression of 1968 gratitude. The reporting went around the table in the rector's office as the deans spoke proudly of their current plans for continued progress.

Several informal conversations dealt with the question "What, after this interval, seems to have been the impact on IPB of the affiliation with the Kentucky team?"

There was the physical presence of a faculty in each of six colleges with department heads, deans and associate deans, the rector and associate rectors; almost all had studied in the United States as participants in the Kenteam period. The basic staff structure of the institution was intact and available to sponsor further development.

"The chief impact has been on the mind of the professor." Lecturing was "more democratic": students asked questions, quizzes were used. The two-semester system was an important change. Laboratory learning and field experiences were the rule; textbooks were used to the extent available, but there were still desperate book shortages. Enrollment was selective and limited. Students were encouraged to organize "for science" and not only politically. The curriculum which formerly required a major and two minors now included a major, a minor, and several electives. Students knew their grades and could check their own progress. The agricultural curriculum was being divided into tracks: one in general agriculture, from which students would go directly into field employment, and another in agricultural science, which should lead to further professional development, doctoral study, service and research in educational institutions. The forestry building at Darmaga had been completed at last and would be occupied within weeks. The details of these comments accumulated without benefit of systematic survey and only incidentally within the rounds of social discourse.

From the time of its founding in 1963 by the merger and redivision of the two faculties in which Kenteam had been helping with preparatory development since 1957, the Agricultural University at Bogor was five years old and had lived as long without as with the guest members of Kenteam. The structure of its colleges and departments was intact, though weakened by lapses due to the part-time application of its energies. The numbers of students and staff had grown, the basic layout of courses had been maintained, there had been further development of governing rules and regulations. There were more college buildings and faculty houses, many of the innovations in teaching had become standard practice. IPB had survived but was consuming both its human and its physical capital in the absence of economic support and technical assistance. The institutional prerequisites to survival were present, and threats were more environmental than internal. By the survival indicator of institutionalization, IPB was a "going concern" and was demonstrating resistance if not resilience in its reaction to adversity.

The Agricultural University's qualities in its third year of berdikari also passed the intrinsic value test. Its recognition as a university-level institute, autonomous within allowable limits of law and ministry regulations, was unquestioned. In no quarter was it challenged any longer whether the separation of the two founding faculties from the University of Indonesia had been warranted. Within the degrees of freedom permitted by the Ministry of Higher Education, IPB selected its own personnel, arranged its own internal organization and its own academic program. Its influence on Indonesian agricultural development was evident in the leadership positions its alumni held, and in its almost proprietorial participation in national efforts to increase food production, especially through the highly regarded Bimas procedures. The weak elements in the test of IPB's intrinsic value lay in its limited ability to attract financial support and in the slow-down of its staffing, both in number and in quality of technical competence. This incapacity was not a special debility of IPB, however. It resulted from deprivation in the whole economy and was beyond the control of any single institution.

The acceptance of IPB's position as a setter-and-bearer of standards was a measure of normative spread. The curriculum and pattern of internal organization at IPB were models for all the new starts in higher education in the agricultural sciences throughout Indonesia. The 1963–1964 concept of IPB

249

as a "feeder" or "mother" (induk) institution had taken a new name—development institution (fakultas pembina). The commitment of IPB to the application of science and to community service (pengabdian masjarakat) had been confirmed by the ministry as national policy. There had been a loss of innovative energy at IPB, and this was a consequence of weaknesses noted above. The creative spark was not dead but it could only smoulder. The urge to innovate was present but possibilities of any forward thrusts had dwindled; the two deficiencies were in general support and in resources to raise levels of competence.

The conclusion, therefore, arrived at by a combination of "rule of thumb" and reference to concepts in a formal outline, is that the institution-building process at Bogor had succeeded in part and was capable of resumption, with ultimate achievement of the full goal if the external restraints of a struggling economy could be removed and support restored.

This summary has dealt mainly with consequences in Indonesia and only briefly with findings that relate to the American involvement. One conclusion of interest to all parties involved may now be stated: Nine years are not enough! It takes longer—probably many years longer—to build an institution which has capacity for self-regeneration and independent growth.

A conclusion of particular importance on the American side is that involvement of the University of Kentucky in the developmental exercise was not organized in sufficient depth, was not incorporated adequately within the total educational program of the university, and was too much of an adjunctive effort. Another conclusion of special concern to AID and American universities is that the contract, dividing responsibility between bureaucrats and educators, did not distinguish well enough their respective competencies.

A final conclusion—of equal interest to all parties in the program and to the future of institution-building effort anywhere—is that the need for cross-cultural effort was urgent enough and the concern of all parties involved was compelling enough to overcome some of the intervening difficulties partially and to overcome others completely. The result was substantial progress toward basic developmental goals.

When American technical assistance in higher agricultural education was resumed in Indonesia in 1969 the affiliation with the University of Kentucky was not revived. There had been deterioration of materiel, erosion of skill, and retrogression in technical competence in the post-coup, post-Kenteam period. Uncertain what the American stance would be against such problems, USAID tested the temper of the Indonesian times by long delay and finally decided to proffer some renewal of aid. The resumption of AID activity in higher agricultural education in Indonesia was preceded by the work of a survey team sent to Djakarta in January 1968. Seemingly preferring a "new slate" approach, AID studiously avoided contact with the University of Kentucky. Initially Kenteam veterans, by wishful thinking later recognized as unrealistic, were puzzled that they were neither included nor consulted. Could it be that IPB had "had enough" of Kentucky? Could it be that Kentucky had lost favor with AID? On inquiry, UK was told that AID wished to be certain of an unbiased review to be undertaken by disinterested parties who could not be suspected of seeking special consideration for their own institutions. It was known to AID that UK would like to resume work in Indonesia and more or less expected to be called back into service when conditions were auspicious. The survey team report was dated January 30, 1968, but was not officially submitted and discussed in AID/Washington until mid-March. There was still no approach to Kentucky, however; university representatives eventually learned in conversations, but were never officially informed, of the attention that was being given to plans for Indonesia. Late in 1968 AID made a contract with a consortium of midwestern universities[5] and began the implementation of the survey team recommendations.

To the Kenteam veterans who visited Indonesia briefly in May and June 1968, it seemed possible to explain several elements in the complex of reasons for disregard of the University of Kentucky while formulating new plans for techni-

[5] MUCIA: The Midwestern Universities Consortium for International Assistance (the Universities of Illinois, Indiana, and Wisconsin, and Michigan State University). The University of Minnesota became a member in 1969.

cal assistance to higher agricultural education in Indonesia. One reason for the nonapproach to Kentucky was probably an American attitude paralleling the Indonesian distinction between the "old order" (of the Sukarno period) and the "new order" (of the Suharto period, after March 11, 1966). The attitude required a purge of the older order, and it could be noted that both embassy and AID staffs in Djakarta were constituted almost entirely of new personnel. None of the university affiliations formerly sponsored by USAID had been reopened. It was intended that AID activities in Indonesia would constitute a completely new program and would be staffed to assure "low visibility" of American personnel. Only the Ford Foundation had decided to capitalize directly on past experience, expertise, and established goodwill by reviving some of the projects it had sponsored earlier.

Another set of contributing factors included certain disaffections with the university on the part of both AID and Indonesian officials at IPB and in the Ministry of Education in Djakarta and the Indonesian embassy in Washington. It should be noted here that, although this book is about Kenteam at Bogor, the attitudes of various agencies toward the University of Kentucky had roots also in their experience with the work of UK at the Institut Teknologi in Bandung. A major accusation was that certain IPB participants had overstayed their leave in the United States, allegedly with University of Kentucky complicity or at least by UK oversight. University personnel had responded by disclaiming status as an agency of deportation and by claiming that the responsibility had been actually with AID and the Immigration Service. The charge of UK negligence in the matter, though never made other than informally, was not withdrawn and there was residual ill-will between some AID personnel and the university.

There was in addition a heritage of minor but cumulative AID discontent over a list of incidents and relationships, including, for example, controversy during negotiations for the extension of participants, the reluctance of some Kenteam wives to accept evacuation on order of the ambassador on November 1, 1965, the delay of a few vexatious Kenteam members in leaving Indonesia by the official withdrawal date, and the chronic Kenteam insistence on autonomy within contract limits from AID's interventional suggestions on programming, staffing, commodity purchasing and participant training. Another possible deterrence was the auditor's challenge of several University of Kentucky expenditures in the Indonesian program. Prolonged review resulted, however, in substantiating most of the university's claims to the legitimacy of reimbursements and in absolving the university of any suspicion of maladministration. Perhaps there was even a political consideration, the requirement that sponsorship of federally-financed programs are to be passed around among the states, and that Kentucky had, after all, had a ten-year chance. Shouldn't the chance go to another institution now, in another state?

One factor, no doubt of major weight, was USAID's memory that only 40 percent of Kenteam's personnel in Bogor had been bona fide university faculty members. Could the university, against this record, be counted on for capability to conduct another effort in agricultural technical assistance after becoming committed in 1967 to maintain a group of ten or more members serving the development of an agricultural center in northeast Thailand? Could this not be related also to awareness in both AID and IPB that special competence in forestry, fisheries, sugar cane, and other tropical crops would not be readily found in Kentucky?

Another factor—perhaps the most important, yet the most subtle to identify and explain—was a mutually felt embarrassment and self-consciousness which lingered in the feeling of both Indonesians and Americans after Kenteam's 1966 departure in an atmosphere of high tension, some anger, and considerable fear. There was also the likelihood that difficulty with the University of Kentucky graduate school might have left a scar or two, even in the hearts of highly placed Indonesian educators. There were complex feelings of regret and fervent though unexpressed hopes that the past might be forgotten. Indonesians who had been forced by the politics of the moment to avoid Kenteam in 1965 and 1966, or even to connive against it, were embarrassed, afraid Kentuckians might remember in the act of returning and might harbor resentments, afraid that IPB as an institution would find it unpleasant to reverse face—having moved in 1965 to send Kenteam away—now to ask that Kenteam come back.

An urgent need for resumption of technical assistance had come to be acknowledged by IPB, but if resumed let it be a clean start, perhaps by affiliation with a grouping of American universities, drawing on the manpower of several institutions rather than only one, with new names and new faces. It was quite clear that IPB made no insistent request of AID that the University of Kentucky be brought again into affiliation with them. But had it been the experience of Kenteam that IPB would make insistent requests? Later, when Americans were asked why Kentucky wasn't invited back, they said Indonesians didn't want this; when Indonesians were asked, they said Americans didn't want it. It must be concluded that both Indonesians and Americans preferred some other arrangement.

Across Hospital Street, Bogor, from the main building of the Agricultural University is the row of faculty houses where Kenteam lived from 1957 until 1966. IPB's families live there now. The gardens are prettier. There are still the coconut, papaya, banana, bamboo, hibiscus, and Christmas roses all year long. The flowers and trees in the front yards have grown and are more luxuriant; grounds around the main building of IPB are being landscaped. Kenteam had trouble with trash—cans would get thrown around; papers would blow; piles of refuse would accumulate. It's cleaner now, and an IPB neighbor laughs, "You have to have enough to eat before you can have much waste!"

It's not recorded that any Kenwife noticed, but a happy IPB wife says, "It's the nicest thing—we can hear in the night the first cry of the newborn baby in the hospital behind us." And there are more ghosts in the neighborhood now than when Kenteam lived there; or maybe Kenteam didn't notice them so much.

Appendix A

Abbreviations Used in This Book

ADC   Agricultural Development Council, Inc.

AID   Agency for International Development

APO   Army Post Office

CDC   Center for Developmental Change, University of Kentucky

CIC   Committee on Institutional Cooperation

Drs.   Doctorandus degree

HSI   Himpunan Sardjana Indonesia [Union of Indonesian Scholars] (university graduates)

ICA   International Cooperation Administration

IPB   Institut Pertanian Bogor [Agricultural University, Bogor]

Ir.   Insinjur degree

ITB   Institut Teknologi Bandung [Institute of Technology, Bandung]

KCT   Kentucky Contract Team (Kenteam)

KRF   Kentucky Research Foundation

OPSSR   Operation Self-sufficiency in Rice

PKI   Partai Kommunis Indonesia [Communist Party, Indonesia]

SSBM   Self-Sufficiency in Bahan Makanan [food materials]

UI   Universitas Indonesia [University of Indonesia]

UK   University of Kentucky

UKOOP   University of Kentucky Office of Overseas Programs

USIS   United States Information Service

Appendix B

Selected characteristics of Kenteam or of Kenteam's work. The table answers the question: What percentage of the Kenteam and IPB respondents agreed that most Kenteam members had the specified characteristics?

	PERCENTAGE	
CHARACTERISTIC	Kenteam	IPB
Had a good "mix" of qualities	92	67
Were technically well-enough qualified	89	87
Recognized their departmental roles	80	63
Had enough organizational skill	75	25
Were not overly committed to American educational methods	72	56
Pushed hard enough	69	47
Had enough political tact	61	21
Understood IPB goals	58	30
Were not "too direct" in making suggestions	58	36
Introduced improved teaching methods	58	80
Travelled enough in Indonesia	56	51
Were sufficiently aloof from factions	56	26
Introduced improved methods of student evaluation	54	67
Had a "sense of mission"	53	40
Had empathy	50	47
Presence in IPB meetings had inhibitory effect	45	38
Abilities were sufficiently utilized	45	30

TABLE II. FEATURES OF THE IPB IMAGES, SELF AND ASCRIBED

Selected characteristics of IPB *or of* IPB's *program. The table answers the question: What percentage of* IPB *and Kenteam respondents agreed that* IPB *had the specified characteristics?*

| | PERCENTAGE | |
CHARACTERISTIC	IPB	Kenteam
Decisions reached by unanimity:		
in departments	87	37
in faculties	77	33
in IPB	70	28
"Equal treatment" principle governed: assignments of Kenteam to		
departments	37	31
procurement of commodities	47	64
selection of participants	63	53
extension of participants' tours	40	32
Practice of advanced planning well enough developed:		
at departmental level	74	36
at faculty level	70	28
at institute level	57	31
Capacity for self-sustaining growth:		
in general	83	86
in my field:		
in undergraduate teaching	73	75
in graduate teaching	76	41
in research	76	44
in extension	70	36
Curricula progressed sufficiently:		
undergraduate	70	50
graduate	53	17
Students' lack of agricultural experience forced Kenteam to modify teaching	39	33
Undergraduate years were overburdened with courses	60	56
Fifth year sufficiently stresses "discovery" (vs. "memory")	50	25
Progressed greatly in library and use of books	90	56

TABLE III. PROBLEMS OF OVERSEAS WIVES

Number of Kenwives reporting feelings of frustration on arrival at Bogor and later, in 18 designated areas of experience in order of number reporting after six months in Indonesia.

AREA OF EXPERIENCE	NUMBER WHO REPORTED FEELINGS OF FRUSTRATION			
	First six weeks in Bogor	After six months	After one year	At time of departure from Bogor
News	22	24	26	23
Traffic	25	23	21	19
Poverty	20	20	20	20
Health	20	20	20	18
Language	26	19	18	15
Standard of living	18	19	17	13
Americans overseas	11	19	19	18
Cleanliness & sanitation	18	17	16	14
Servants	14	17	20	12
Time	17	16	10	9
Etiquette	13	14	10	7
Population	15	14	12	11
Revolution	13	14	17	17
Values	16	14	13	13
Money	15	12	13	11
Food	16	11	4	3
Climate	10	11	9	8
Culture	9	6	6	4

Index

Black Muslims, 21
Bogor, Indonesia, *mentioned passim;* mayor, 41
British embassy, 22, 59; evacuation of families, 22
Brussels, Belgium, 11
Bunker, Ambassador Ellsworth, 21

California, University of, Los Angeles, 6, 28, 42
Calvert graded school materials, 199
campus, concept of, 73, 74
campus coordinator. *See* Kentucky, University of
Center for Developmental Change (CDC), ix, xi, xvii, 7, 142
Center for International Studies, Cornell University, 240
Cereal Crops Research Institute, 93
change agents, xi, 244; Kenteam as, 42. *See also* developmental change
China, 21
Chinese Indonesians, 20, 206; shops of attacked, 21, 50; foods and cooking of, 206, 235, 236, 237; porcelain, 235
Church World Service, 14
CIC. *See* Committee on Institutional Cooperation
civil defense, 22
college. *See* faculty
Committee on Institutional Cooperation (CIC): CIC-AID rural development research project, xiii, xvii, 62, 239; recommendation of, xiii, xiv, 149, 239; and AID, 243
commodity program. *See* purchasing program
communication: etiquette of, 33, 34; problems of, 34, 38, 39, 40, 52; Indonesian speech customs in, 35; techniques of, 37; rumor in, 37, 38; among partners in Kenteam project, 62-65; ground rules for, 62; physical means of, 63, 209; bureaucratic, 63; personal vs. impersonal, 64; and status structure, 64

communism, 5, 14, 16, 18, 23, 178
community service, 12, 250
confrontation with Malaysia (Konfrontasi), 22
consensus. *See* decision-making
contracts. *See* AID; Kenteam, ground rules; Kentucky, University of; technical assistance
Council on Economic and Cultural Affairs. *See* Agricultural Development Council, Inc.
coup-attempt of September 30, 1965, 28; post-coup period, xv, 151
cross-cultural experience, orientation for, 21, 45, 119, 145-48
Cuban crisis, 21, 50
cultural shock, 200, 238. *See also* Kenwives
culture, Indonesian: cultural layers in Java, 4; spirit world, 4, 5, 36, 238; dukun (shaman), 36, 228; selametans, 36; systems of causation, 36; decision by consensus, 37; values, 43, 44, 53; culture differences, 44, 45; limited good, doctrine of the, 44, 98, 113, 180, 181; mutual aid (gotong rojong), 44, 98, 180; hierarchies and rank, 45; sila posture, 76; meaning of time, 82, 228; strain toward consistency, 132; father as decision-maker, 180; equality principle, 181; American responses to, 197-238; sembah gesture, 231. *See also* decision-making; Islam; Kenwives; rituals, Indonesian
Curators, Board of. *See* Dewan Curator
curricula. *See* IPB, curricula

Darmaga, Indonesia, 15, 16, 73, 247, 248
decision-making: by consensus, 37; types of, compared, 179-80
degrees. *See* academic degrees
Department of Higher Education and Science. *See* Education, Ministry of

developmental change: strategy of, xi; change agents, xi, 42, 244; group approach, xii; economic development, 9; motivation for, 9; defined, 26; developmental services, 106-8; institution for (fakultas pembina), 250

Dewan Curator (Board of Curators), 18, 41, 107

Dickey, Frank G., 140

Djakarta, Indonesia. *Mentioned passim*

djam karet, 228

Djogjakarta silver, 235

Djuanda, Dr., viii

Donovan, Herman L., 140

drama: wayang, 4, 205, 235

Dutch: faculty at IPB, 9; relationships with Indonesia, 18, 19; withdrawal from Bogor, 20; educational system, 80, 109; goals for IPB, 159

Dutch East India Company, 5

Economic Cooperation Administration, U.S., vii, 6

economic stabilization, 22

Education, Ministry of, 6, 15, 23, 28, 51, 85, 105, 108

education in Indonesia: levels of, 9; community (social), 9; history of development in, 9; and national development, 9; basic law on universities, 11; study of needs, 23, 51-52; Dutch, 80; Indonesianization of, 183. *See also* Education, Ministry of; IPB, curricula; universities

eight-year plan, 20

equipment. *See* purchasing program

Esman, Milton J., 240n

ethnic groups in Indonesia, 4

Europe: postwar recovery of, 5; influence on Indonesian culture, 4, 5

experiment stations, 93. *See also* institutes, tropical; research, agricultural

extension education. *See* IPB

extension services. *See* agricultural services; people's agriculture

faculty (college): medical, 6; agricultural, 9, 10; veterinary, 10, 41; forestry, 12, 74; fisheries, 43, 94, 95; agricultural technology and mechanization, 86. *See also* IPB

fakultas induk, 105, 250

fakultas pembina, 105

family planning, 232

fisheries: college and curriculum at IPB, 43, 84, 94, 95

food: shortages, 14; 1968 national workshop on, 247

Ford Foundation, xvii, 252; withdrawal of, 27

foreign aid, university participation in, 6

Forest and Livestock Services, 107, 190

forestry: faculty of at IPB, 12, 74; curriculum, 84; field work in, 97

frustration. *See* Kenwives

Fulbright sponsorship, xvi

Gadjah Mada, University of, 10, 28, 109

gamelan music, 4, 205-6

Gardner Report, 149

Geertz, Clifford, 45

generation of '66 (generasi '66), 16

Germany, 76

Goodyear Tire factory, Bogor, 26

gotong rojong, 44, 98, 180

grading system. *See* IPB, teaching practices

Great Garden, Bogor, 3

Greece, xv, 5

Green, Ambassador Marshall, 28

group approach: as a strategy of change, xii

Guatemala, vii, xi, 140

guerrilla warfare, 5

guest professor. *See* host-guest model

guided study. *See* IPB, curricula

Gujarat, India, 5

Hadiwidjaja, Tojib, viii, 10, 11, 23, 81
Hansip (civil defense), 22
Higher Education Week, 178
Himpunan Sardjana Indonesia, 26
Hindu culture, 4, 5
home economics, xv, 7, 96
host-guest model, 32; host institution, xv, 7; host colleagues, 7, 244; guest professors, xiv, xv, 7, 10, 32, 244
HSI. *See* Himpunan Sardjana Indonesia

ICA. *See* International Cooperation Administration
Illinois, University of, 251
illiteracy, 9
Immigration Service, U.S., 252
India, xvi; war with Pakistan, 21
Indiana, University of, 6n, 251
Indonesia, Republic of, 18
Indonesia, University of. *Mentioned passim*
Indonesian-American ladies' group, 50
Indonesian culture. *See* culture, Indonesian
Indonesian Embassy, Washington, D.C., 252
Indonesian foods, 27, 236
Indonesian goals. *See* berdikari; IPB
Indonesian identity, recognition of, 49
Indonesian independence, proclamation of, 3, 18
Indonesianization process, 188
Indonesian Revolution, 3, 178. *See also* "situation, the"
Indonesian Science Institute, 247
Indonesian-U.S. relations: anti-American sentiment, 20; estrangement and American withdrawal, 27. *See also* "situation, the"
Indonesia-Peking axis, 26
Institute of Agricultural Sciences. *See* IPB
institutes, tropical, 11, 100
institution-building: study of, xi,

xii; role models at Bogor, 53; Pittsburgh model, 149, 240, 244, 245; defined, 238, 240; concepts applied to Kenteam case, 240-49; usefulness of concept, 245-46; time requirements for, 250. *See also* Agency for International Development; IPB
Institut Teknologi Bandung (ITB), viii, 6, 10, 252
International Cooperation Administration (ICA), vii, 6, 7, 10, 140. *See also* Agency for International Development
IPB (Institut Pertanian Bogor, the Agricultural University at Bogor), ix, 4, 6, 12, 14; extension education at, 7, 17, 83, 104; as host institution, 9; first rector of, 11; organization of, 12-13, 17, 43; presidium, 12, 25; senates, 13; staff, 13-14, 105; financing of, 14-15; alumni, 18; difficulties in use of English, 32, 91, 92; agreement and differences within, 40, 43, 47; educational issues at, 43; flying professors, 105; and academies and institutes, 105-6; development of educational programs, 108-10; innovation at, 109; goals of, 159, 241; self- and ascribed images of, 178-96, 257; symbol of, 178; identity of, 179; consensus and unanimity within, 180-81; planning at, 182; capacity for growth, 182-85; books and library, 192-94; status in 1966, 195-96; status in 1968, 248-49; as a mother (induk) institution, 250. *See also* Kenteam; participant program; postgraduate work; research, agricultural
—curricula: fields included, 6; development of, 13, 81, 83, 85, 86; guided study, 43, 84, 98; academic calendar, 80, 82; time schedule, 82; catalog (pedoman), 82; testimonium, 83, 84; departmental unity, 84; free

to American educational
methods, 155-56; aggressiveness,
156-57; political skill, 157-58;
understanding of goals, 158-60;
empathy, 167; attitudes, 176;
and the Kenteam project, 241;
impact as observed in 1968,
248-49; housing of, 3, 71, 254;
percentage from University of
Kentucky, 253
—relationships of, 176; with IPB,
30, 32; extent of acquaintance,
39-40; restraint as a policy in,
50; tension at time of withdrawal, 52; host-guest, 53; role
ambiguities in, 53; as a group,
56; as a team, 56, 243, 244;
with AID, 135-40 passim;
solidarity and alienation in, 138-
40; with the University of
Kentucky, 140-50 passim;
neutrality as a policy in, 164-65;
inhibitory effect of, 167-68
Kentuckians, ethnocentrism of, 8
Kentucky, University of: project in
Guatemala, vii, xi, 140; Office
of Overseas Programs, ix, 7, 142;
campus coordinator of Kenteam
project, 54, 142-45; qualifications
for technical cooperation, 140-
41; experience in Appalachians,
140; relationship with Kenteam,
140-50; college of agriculture,
142; criticism of by AID 252;
post-contract attitudes toward,
253-54. See also Agency for
International Development
—AID contracts for work in
Indonesia: at Bandung, viii, 21,
27, 28, 250; at Bogor, xiv;
budget of, 15; extension of, 20;
campus coordinator for, 22, 142;
problems of interpreting, 70;
nature of, 149; understanding
of, 149. See also participant
program; purchasing program
Kentucky Contract Team. See
Kenteam
Kentucky Research Foundation
(KRF), ix, 7, 141
Kenwives (wives of Kenteam

members), xv, 23, 26; evacuation
of, 28, 252; survey of, 197, 198;
motivation of, 198; problems of
preparation by, 198-200;
friendship of, 201, 202, 203;
activities of, 203, 204; teaching
by, 204; cultural participation
of, 204, 205; regrets of, 205;
Bogor-Indonesian women's
meetings, 206; association with
Bogor Mothers, 206
—areas of shock and frustration:
news, 208, 209; traffic, 209-11;
poverty, 211, 212; health, 213,
214; language, 214-17; revolutionary atmosphere, 217, 218;
standards of living, 219, 220;
housing, 219-23; other
Americans overseas, 223, 224;
cleanliness and sanitation, 224,
225; experience with servants,
225-28; time values, 228-30;
etiquette, 230-33; population
problems, 232; cultural values,
233; money, 234, 235; food, 235-
37; climate, 237, 238
Kisaran, Sumatra, 94
Konfrontasi, 22
KRF. See Kentucky Research
Foundation

land-grant colleges, 8, 109, 149,
159
Laos, 21
Lebaran, 203, 212
Lembaga Ilmu Pengetahuan
Indonesia, 247
limited good, doctrine of the, 44,
98, 113, 180, 181
Lumumba, Patrice, 20, 50
lurah (village head), 23

Madjapahit, Kingdom of, 5
Madura, 4, 93
Mahabharata, 5
marine fisheries. See fisheries
Marshall Plan, 5
Mataram, Kingdom of, 5
MATS. See Military Technical
Assistance Group
Medan, Indonesia, 25

This book has been set in
W. A. Dwiggins' Linotype Caledonia
with display set in ATF Bulmer;
printing by the Printing Dept.
of the University of Kentucky;
design by James Wageman.

www.ingramcontent.com/pod-product-compliance
Lightning Source LLC
Chambersburg PA
CBHW020340270326
41926CB00007B/251